I0469425

August 2012

SOLAR ENERGY

Federal Initiatives Overlap but Take Measures to Avoid Duplication

Accountability ★ Integrity ★ Reliability

GAO-12-843

GAO
Accountability * Integrity * Reliability

Highlights

Highlights of GAO-12-843, a report to
congressional requesters

SOLAR ENERGY

Federal Initiatives Overlap but Take Measures to Avoid Duplication

Why GAO Did This Study

The United States has abundant solar energy resources and solar, along with wind, offers the greatest energy and power potential among all currently available domestic renewable resources. In February 2012, GAO reported that 23 federal agencies had implemented nearly 700 renewable energy initiatives in fiscal year 2010—including initiatives that supported solar energy technologies (GAO-12-260). The existence of such initiatives at multiple agencies raised questions about the potential for duplication, which can occur when multiple initiatives support the same technology advancement activities and technologies, direct funding to the same recipients, and have the same goals.

GAO was asked to identify (1) solar-related initiatives supported by federal agencies in fiscal years 2010 and 2011 and key characteristics of those initiatives and (2) the extent of fragmentation, overlap, and duplication, if any, of federal solar-related initiatives, as well as the extent of any coordination among these initiatives. GAO reviewed its previous work and interviewed officials at each of the agencies identified as having federal solar initiatives active in fiscal years 2010 and 2011. GAO developed a questionnaire and administered it to officials involved in each initiative to collect information on: initiative goals, technology advancement activities, funding obligations, number of projects, and coordination activities.

This report contains no recommendations. In response to the draft report, USDA generally agreed with the findings, while the other agencies had no comments.

View GAO-12-843. For more information, contact Frank Rusco at (202) 512-3841 or ruscof@gao.gov.

What GAO Found

Sixty-five solar-related initiatives with a variety of key characteristics were supported by six federal agencies (see table). Over half of these 65 initiatives supported solar projects exclusively; the remaining initiatives supported solar and other renewable energy technologies. The 65 initiatives exhibited a variety of key characteristics, including multiple technology advancement activities ranging from basic research to commercialization by providing funding to various types of recipients including universities, industry, and federal laboratories and researchers, primarily through grants and contracts. Agency officials reported that they obligated about $2.6 billion for the solar projects in these initiatives in fiscal years 2010 and 2011, an amount higher than in previous years, in part, because of additional funding from the 2009 American Recovery and Reinvestment Act.

Number of Federal Initiatives That Supported Solar Energy Technology in Fiscal Years 2010 and 2011, by Agency and Total Obligations

Agency	Number of solar initiatives	Total obligations
Department of Energy (DOE)	20	$2,342,765,827
Department of Defense (DOD)	27	154,637,296
National Science Foundation (NSF)	7	48,761,556
National Aeronautics and Space Administration (NASA)	7	34,818,345
U.S. Department of Agriculture (USDA)	3	577,247
Environmental Protection Agency (EPA)	1	200,000
Total	65	$2,581,760,271

Source: GAO analysis of agency-provided data.

The 65 solar-related initiatives are fragmented across six agencies and overlap to some degree in their key characteristics, but most agency officials reported coordination efforts to avoid duplication. The initiatives are fragmented in that they are implemented by various offices across the six agencies and address the same broad areas of national need. However, the agencies tailor their initiatives to meet their specific missions, such as DOD's energy security mission and NASA's space exploration mission. Many of the initiatives overlapped with at least one other initiative in the technology advancement activity, technology type, funding recipient, or goal. However, GAO found no clear instances of duplicative initiatives. Furthermore, officials at 57 of the 65 initiatives (88 percent) indicated that they coordinated in some way with other solar-related initiatives, including both within their own agencies and with other agencies. Such coordination may reduce the risk of duplication. Moreover, 59 of the 65 initiatives (91 percent) require applicants to disclose other federal sources of funding on their applications to help ensure that they do not receive duplicative funding.

Contents

Figures

Abbreviations

ARPA-E	Advanced Research Projects Agency-Energy
DARPA	Defense Advanced Research Projects Agency
DOD	Department of Defense
DOE	Department of Energy
EERE	Office of Energy Efficiency and Renewable Energy
EIA	Energy Information Administration
EPA	Environmental Protection Agency
IAPG	Interagency Advanced Power Group
NASA	National Aeronautics and Space Administration
NNI	National Nanotechnology Initiative
NSF	National Science Foundation
OMB	Office of Management and Budget
R&D	research and development
Recovery Act	American Recovery and Reinvestment Act of 2009
SETP	Solar Energy Technologies Program
TRL	technology readiness level
USDA	United States Department of Agriculture

United States Government Accountability Office
Washington, DC 20548

August 30, 2012

The Honorable Ralph M. Hall
Chairman
Committee on Science, Space, and Technology
House of Representatives

The Honorable Andy Harris
Chairman
Subcommittee on Energy and Environment
Committee on Science, Space, and Technology
House of Representatives

The United States has abundant solar energy resources and solar, along with wind, offers the greatest energy and power potential among all commercial domestic renewable resources currently available, according to the National Academies of Science. While helping meet the nation's energy needs, solar sources of electricity could also offer substantial environmental benefits, such as a reduction in greenhouse gas emissions, over conventional electricity generation. However, solar and wind energy as sources of electricity at the utility scale also face numerous challenges related to their intermittent availability and higher costs compared with traditional energy sources. The Department of Energy's (DOE) Energy Information Administration, an independent statistical and analytical agency, reports that although solar energy accounted for only 1 percent of all renewable energy electricity consumed in the United States in 2010, solar energy use increased by about 60 percent from 2006 to 2010. DOE has also embarked on a comprehensive effort to reduce the cost of generating electricity using solar power.

Beyond supplying electricity to the existing infrastructure or grid (i.e., the electricity networks that carry electricity from the plants where it is generated to consumers, including wires, substations, and transformers), solar energy technologies have extensive defense and space applications. For example, the Department of Defense (DOD) has undertaken efforts to develop technologies that allow it to use solar energy in operational environments, such as the conflicts in Iraq and Afghanistan, to help reduce its reliance on conventional fuel, which is very difficult and costly to deliver and store. In addition, the National Aeronautics and Space Administration (NASA) is working to develop high-power solar electric propulsion systems to provide low-cost abundant power for deep-space missions. As the availability and utilization of solar energy technologies have increased, so too has federal support to some of the agencies addressing challenges associated with further expansion of solar energy as an electricity source and to meet other mission-specific goals.

In February 2012, we reported that 23 federal agencies had implemented nearly 700 renewable energy initiatives in fiscal year 2010—and a number of these initiatives supported solar energy technologies.[1] The existence of such initiatives at multiple agencies has raised questions about the potential for duplication that, in this context, occurs when multiple initiatives support the same technology advancement activities and technologies, direct funding to the same recipients, and have the same goals. As we have previously reported, unnecessary duplication can potentially result from fragmentation and overlap among government programs.[2] Fragmentation occurs when more than one federal agency, or more than one organization within an agency, is involved in the same broad area of national need. For the purposes of this report, overlap occurs when multiple initiatives support similar technology advancement activities, similar technologies, similar funding recipients, or have similar goals.

In this context, you asked us to review federal initiatives that supported solar energy technology—which we termed "solar-related initiatives." Our objectives were to identify (1) solar-related initiatives supported by federal agencies in fiscal years 2010 and 2011 and key characteristics of those initiatives and (2) the extent of fragmentation, overlap, and duplication, if any, among federal solar-related initiatives, as well as the extent of coordination among these initiatives.

To address these objectives, we collected and analyzed information from our previous work and conducted new work. More specifically, to identify federal solar-related initiatives and to assess the extent of fragmentation, overlap, and duplication among these initiatives, we relied on data from

---

[1]GAO, *Renewable Energy: Federal Agencies Implement Hundreds of Initiatives*, GAO-12-260 (Washington, D.C.: Feb. 27, 2012).

[2]GAO, *Managing for Results: Using the Results Act to Address Mission Fragmentation and Program Overlap*, GAO/AIMD-97-146 (Washington, D.C.: Aug. 29, 1997). For more information on fragmentation, overlap, and duplication in federal programs see GAO, *Opportunities to Reduce Potential Duplication in Government Programs, Save Tax Dollars, and Enhance Revenue*, GAO-11-318SP (Washington, D.C.: Mar. 1, 2011).

GAO-12-843 Solar Energy

our February 2012 report on renewable energy,[3] which identified and collected information from solar-related initiatives active[4] in fiscal year 2010, as well as other renewable energy initiatives. For the purposes of this report, we defined an initiative as a program or group of agency activities serving a similar purpose or function that involved solar energy technologies through a specific emphasis or focus, even if solar energy was only one part of a broader effort. We restricted our analysis to those initiatives that supported research and development (R&D) on, or commercialization of, innovative solar energy technologies.

Because initiatives often supported more than one type of technology advancement activity, some of the initiatives included in this report may also support deployment activities. We eliminated initiatives that focused solely on the deployment of readily available solar energy technologies. Although not covered in this report, a range of federal, state, and other initiatives support such deployment activities. These federal deployment activities facilitate or achieve widespread use of technologies—for example, by federal agencies' procurement of solar energy technologies or by federal tax or other incentives that encourage households and businesses to adopt these technologies. We have ongoing work that looks at some of these deployment activities, including DOE's Loan Guarantee and Advanced Technology Vehicles Manufacturing Loan Programs, and a review of wind energy technology initiatives, including those that support deployment activities. We also recently issued reports on tax incentives for residential energy efficiency[5] and on renewable energy deployment activities.[6]

After using our February 2012 report to identify solar-related initiatives that supported R&D and commercialization, we developed a questionnaire and surveyed cognizant officials at each of the six agencies that had such initiatives to collect additional information on the initiatives' scope and key characteristics—DOD, DOE, the Environmental Protection

---

[3]GAO-12-260.

[4]For the purposes of this report, we defined "active" initiatives as those that were planned or funded or implemented or authorized in the fiscal year described.

[5]GAO, *Energy Conservation and Climate Change: Factors to Consider in the Design of the Nonbusiness Energy Property Credit,* GAO-12-318 (Washington, D.C.: Apr. 2, 2012).

[6]GAO-12-260.

Agency (EPA), NASA, the National Science Foundation (NSF), and the U.S. Department of Agriculture (USDA). Specifically, we asked questions to ascertain the number of projects, funds obligated, type of technology advancement activities supported, recipients funded, funding mechanisms, initiative goals, and efforts to coordinate across other solar-related initiatives. To determine the extent to which agencies coordinated their initiatives, we used survey responses and interviews to identify coordination activities within and among the six agencies. We also asked the agencies to provide us with any additional relevant initiatives that were implemented in fiscal year 2011 that were not on our initial list, and we sent a copy of the questionnaire to officials representing those initiatives as well. (For further information on our questionnaire, see app. I. For a copy of our questionnaire, see app. III.) We then analyzed the information collected for the February 2012 report and from our questionnaire for this report to determine the extent to which we could identify fragmentation, overlap, and duplication of solar-related initiatives. In addition, we interviewed agency officials responsible for solar-related initiatives for follow-up information as needed.

We conducted this performance audit from September 2011 to August 2012 in accordance with generally accepted government auditing standards. Those standards require that we plan and perform the audit to obtain sufficient, appropriate evidence to provide a reasonable basis for our findings and conclusions based on our audit objectives. We believe that the evidence obtained provides a reasonable basis for our findings and conclusions based on our audit objectives.

Background

Solar energy can be used to heat, cool, and power homes and businesses with a variety of technologies that convert sunlight into usable energy. Examples of solar energy technologies include photovoltaics, concentrated solar power, and solar hot water. Solar cells, also known as photovoltaic cells, convert sunlight directly into electricity. Photovoltaic technologies are used in a variety of applications. They can be found on residential and commercial rooftops to power homes and businesses; utility companies use them for large power stations, and they power space satellites, calculators, and watches. Concentrated solar power uses mirrors or lenses to concentrate sunlight and produce intense heat, which is used to generate electricity via a thermal energy conversion process; for example, by using concentrated sunlight to heat a fluid, boil water with the heated fluid, and channel the resulting steam through a turbine to produce electricity. Most concentrated solar power technologies are designed for utility-scale operations and are connected to the electricity-transmission system. Solar hot water technologies use a collector to absorb and transfer heat from the sun to water, which is stored in a tank until needed. Solar hot water systems can be found in residential and industrial buildings.

Innovation in solar energy technology takes place across a spectrum of activities, which we refer to as technology advancement activities, and which include basic research, applied research, demonstration, and commercialization.[7] For purposes of this report, we defined basic research to include efforts to explore and define scientific or engineering concepts or is conducted to investigate the nature of a subject without targeting any specific technology; applied research includes efforts to develop new scientific or engineering knowledge to create new and improved technologies; demonstration activities include efforts to operate new or improved technologies to collect information on their performance and assess readiness for widespread use; and commercialization efforts transition technologies to commercial applications by bridging the gap between research and demonstration activities and venture capital funding and marketing activities.

Solar energy technology advancement activities are financed through both public and private investment. According to a Congressional Budget Office report, without public investment, the private sector's investment in technology advancement activities is likely to be inefficiently low from society's perspective because firms cannot easily capture the "spillover benefits" that result, particularly at the early stages of developing a technology.[8] In these stages, technology advancement activities can create fundamental knowledge leading to numerous benefits for society

---

[7]We developed definitions that could be applied broadly to make comparisons across agencies and that covered the full spectrum of advancement activities. Federal agencies use various definitions and categories for describing the stages of technology advancement. For example, NASA and DOE use technology readiness level (TRL) categories and definitions to measure and communicate technology readiness for first-of-a-kind technology applications. However, these agencies' TRL categories and definitions are not the same. In addition, the Office of Management and Budget (OMB) provides federal definitions for the terms basic research, applied research, and development in OMB Circular A–11, Section 84—Character Classification (Schedule C), but does not provide definitions for demonstration activities and commercialization activities. During pretests of our questionnaire, agency officials were able to fit their initiatives' activities within our categories and based on our definitions, which validated our use of them.

[8]Congressional Budget Office, *Federal Financial Support for the Development and Production of Fuels and Energy Technologies* (Washington, D.C.: March 2012).

as a whole but not necessarily for the firms that invested in the activities. For example, basic research can create general scientific knowledge that is not itself subject to commercialization but that can lead to multiple applications that private companies can produce and sell. As activities get closer to the commercialization stage, the private sector may increase its support because its return on investment increases.

Six Agencies Supported Sixty-Five Solar-Related Initiatives with a Variety of Key Characteristics

We identified 65 solar-related initiatives with a variety of key characteristics at six federal agencies.[9] Over half of the 65 initiatives supported solar projects exclusively; the remaining initiatives supported solar energy technologies in addition to other renewable energy technologies. The initiatives demonstrated a variety of key characteristics, including focusing on different types of solar technologies and supporting a range of technology advancement activities from basic research to commercialization, with an emphasis on applied research and demonstration activities. Additionally, the initiatives supported several types of funding recipients including universities, industry, nonprofit organizations, and federal labs and researchers, primarily through grants and contracts. Agency officials reported that they obligated around $2.6 billion for the solar projects in these initiatives in fiscal years 2010 and 2011.

Over Half of the Agencies' Initiatives Exclusively Supported Solar Energy Technology

In fiscal years 2010 and 2011, six federal agencies—DOD, DOE, EPA, NASA, NSF, and USDA—undertook 65 initiatives that supported solar energy technology, at least in part. (See app. II for a full list of the initiatives). Of these initiatives, 35 of 65 (54 percent) supported solar projects exclusively and 30 (46 percent) also supported projects that were not solar. For example, in fiscal years 2010 and 2011, DOE's *Solar Energy Technologies Program—Photovoltaic Research and Development* initiative, had 263 projects, all of which focused on solar energy. In contrast, in fiscal years 2010 and 2011, DOE's *Hydrogen and Fuel Research and Development* initiative—which supports wind[10] and other renewable

---

[9]We identified as solar-related those initiatives that supported or could have supported solar technologies, either solely or as part of a larger suite of renewable energy technologies, such as wind or biofuels. For the purposes of this report, we defined an initiative as a program or group of agency activities serving a similar purpose or function that involved solar energy technologies through a specific emphasis or focus, even if solar energy was only one part of a broader effort.

[10]Wind energy is energy produced by the movement of air.

sources that could be used to produce hydrogen—had 209 projects, 26 of which were solar projects. Although initiatives support solar energy technologies, in a given year, they might not support any solar projects. For example, NSF officials noted that the agency funds research across all fields and disciplines of science and engineering and that individual initiatives invite proposals for projects across a broad field of research, which includes solar-related research in addition to other renewable energy research. However, in any given year, NSF may not fund proposals that address solar energy because either no solar proposals were submitted or the submitted solar-related proposals were not deemed meritorious for funding based upon competitive, merit-based reviews.

Although more than half of the agencies' initiatives supported solar energy projects exclusively, the majority of projects supported by all 65 initiatives were not focused on solar. As shown in table 1, of the 4,996 total projects active in fiscal years 2010 and 2011 under the 65 initiatives, 1,506 (30 percent) were solar projects, and 3,490 (70 percent) were not solar projects.

Table 1: Initiatives and Projects by Agency, Fiscal Years 2010 and 2011

Agency	Number of initiatives	Total projects supported	Total solar projects supported	Percentage of projects that were solar
DOD	27	623	293	47%
DOE	20	1,736	906	52%
EPA	1	239	14	6%
NASA	7	1,326	202	15%
NSF	7	349	86	25%
USDA	3	723	5	.7%
Total	**65**	**4,996**	**1,506**	**30%**

Source: GAO analysis of agency-provided data.

GAO-12-843 Solar Energy

Key Characteristics of Solar-Related Initiatives Include Supporting Different Solar Technologies, Technology Advancement Activities, and Types of Funding Recipients

Agencies' solar-related initiatives supported different types of solar energy technologies. According to agency officials responding to our questionnaire, 47 of the 65 initiatives supported photovoltaic technologies, and 18 supported concentrated solar power; some initiatives supported both of these technologies or other solar technologies.[11] For example, NSF's *CHE-DMR-DMS Solar Energy Initiative (SOLAR)*[12] supports both photovoltaic and concentrated solar power technologies, including a project that is developing hybrid organic/inorganic materials to create ultra-low-cost photovoltaic devices and to advance solar concentrating technologies.

These initiatives supported solar energy technologies through multiple technology advancement activities, ranging from basic research to commercialization. As shown in figure 1, five of the six agencies supported at least three of the four technology advancement activities we examined, and four of the six supported all four.

---

[11]Rather than specifying that the initiative specifically supported photovoltaic or concentrating solar power, some agency officials indicated that their initiative supported "solar" technology as a general category. Others indicated that they supported "other" solar technology, such as DOD's *Novel Power Sources* initiative, which supports solar fuels.

[12]According to NSF, the purpose of the *CHE-DMR-DMS Solar Energy Initiative* is to support interdisciplinary efforts by groups of researchers to address the scientific challenges of highly efficient harvesting, conversion, and storage of solar energy. Such groups must include three or more co-Principal Investigators, of whom one must be a researcher in chemistry, a second in materials, and a third in mathematical sciences, in areas supported by the Divisions of Chemistry (CHE), Materials Research (DMR), and Mathematical Sciences (DMS), respectively.

Figure 1: Number of Agencies' Solar-Related Initiatives That Supported Technology Advancement Activities

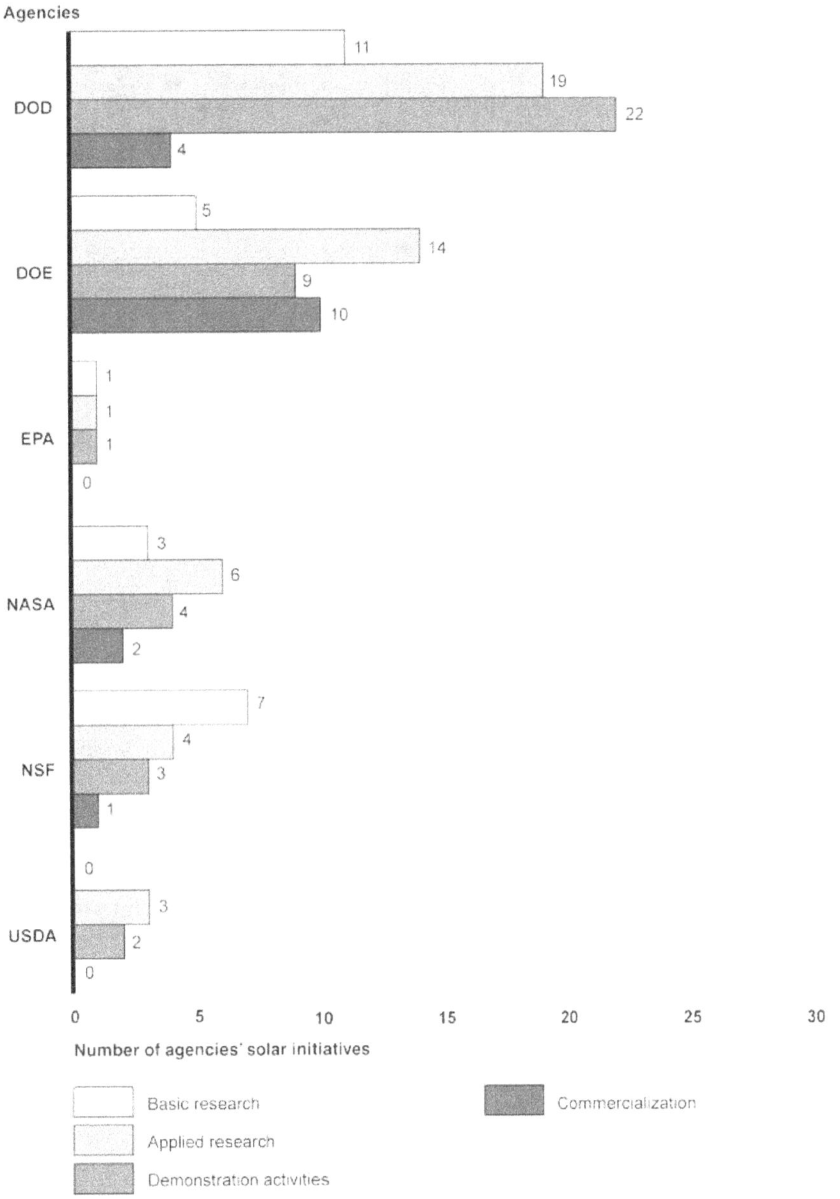

Agencies

Number of agencies' solar initiatives

Legend:
- Basic research
- Applied research
- Demonstration activities
- Commercialization

Source: GAO analysis of agency-provided data

Note: Because agency initiatives often supported more than one technology advancement activity, the sum of initiatives across the agencies in this figure will not total 65, the number of initiatives in our review.

Our analysis showed that of the 65 initiatives, 20 initiatives (31 percent) supported a single type of technology advancement activity; 45 of the initiatives (69 percent) supported more than one type of technology advancement activity; and 4 of those 45 initiatives (6 percent) supported all four. For example, NASA's *Solar Probe Plus Technology Development* initiative—which tests the performance of solar cells in elevated temperature and radiation environments such as near the sun—supported applied research exclusively. In contrast, NASA's *Small Business Innovations Research/Small Business Technology Transfer Research* initiative—which seeks high-technology companies to participate in government-sponsored research and development efforts critical to NASA's mission—supported all four technology advancement activities. The technology advancement activities supported by the initiatives were

- applied research (47 initiatives),

- demonstration (41 initiatives),

- basic research (27 initiatives), and

- commercialization (17 initiatives).

The initiatives supported these technology advancement activities by providing funding to four types of recipients: universities, industry, nonprofit organizations, and federal laboratories and researchers. The initiatives most often supported universities and industry. In many cases, initiatives provided funding to more than one type of recipient. Specifically, our analysis showed that of the 65 initiatives, 23 of the initiatives (35 percent) supported one type of recipient; 21 of the initiatives (32 percent) provided funding to at least two types of recipients; 17 initiatives (26 percent) supported three types; and 4 initiatives (6 percent) supported all four. In two cases, agency officials reported that their initiatives supported "other" types of recipients, which included college students and military installations.

Initiatives often supported a variety of recipient types, but individual agencies more often supported one or two types. As shown in figure 2, DOE's initiatives most often supported federal laboratories and researchers; DOD's most often supported industry recipients; NASA's supported federal laboratories and industry equally; NSF's supported universities exclusively. For example, NASA's *Small Business Innovations Research/Small Business Technology Transfer Research* initiative provided contracts to industry to participate in government-sponsored research and development for advanced photovoltaic technologies to improve efficiency and reliability of solar power for space exploration missions. NSF's *Emerging Frontiers in Research and Innovation* initiative provided grants to universities for, among other purposes, promoting breakthroughs in computational tools and intelligent systems for large-scale energy storage suitable for renewable energy sources such as solar energy.

Figure 2: Number of Agencies' Solar-Related Initiatives That Supported Various Recipient Types

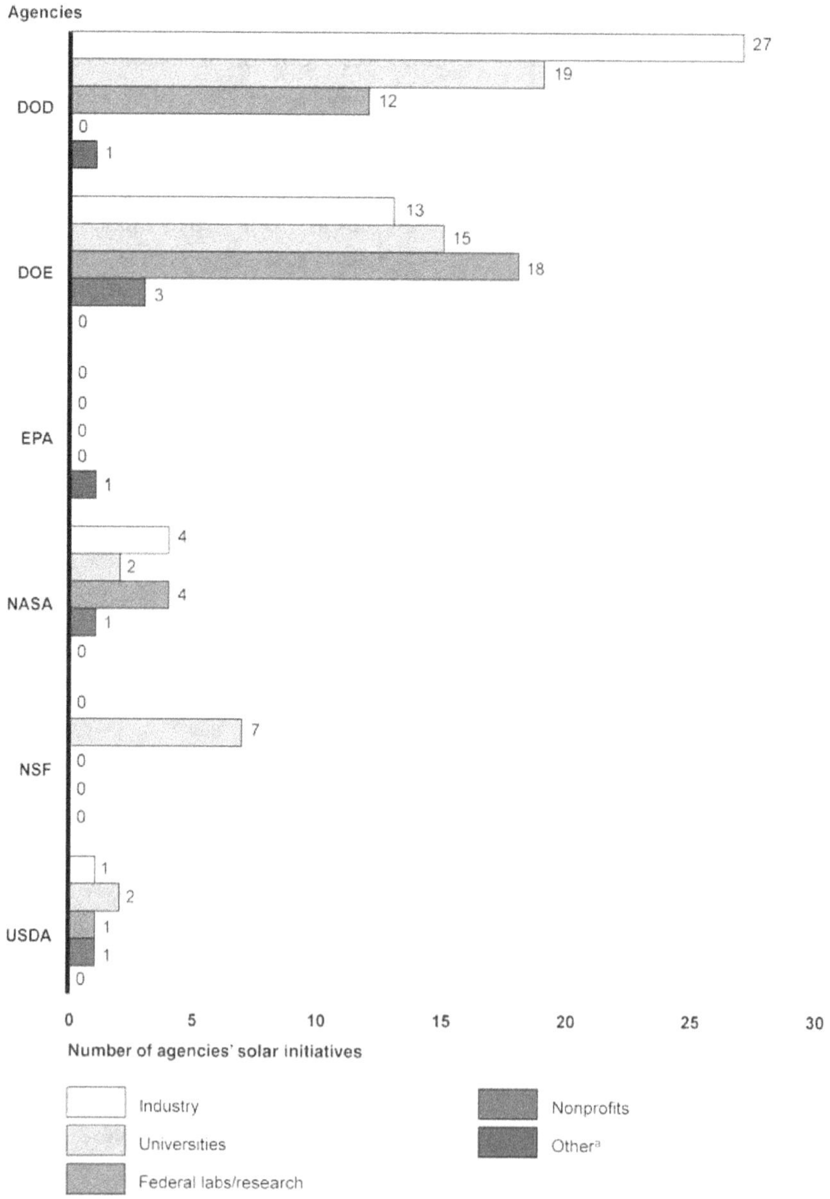

Agencies

Number of agencies' solar initiatives

Legend:
- Industry
- Universities
- Federal labs/research
- Nonprofits
- Other[a]

Source: GAO analysis of agency-provided data.

Note: Agency initiatives often supported more than one recipient type; therefore, the sum of initiatives across the agencies in this figure will not total 65, the number of initiatives in our review.

[a]EPA's *P3: People, Prosperity & the Planet Student Design Competition for Sustainability* initiative supports teams of college students who participate in a competitive grants program in which they propose the research and development of sustainable designs related to water, energy, agriculture, the built environment, or materials and chemicals. DOD's *Smart Power Infrastructure Demonstration for Energy Reliability and Security (SPIDERS) Joint Capability Technology Demonstration* supports military installations to integrate renewable energy and storage in order to reduce the risk of extended electric grid outages.

Federal solar-related initiatives provided funding to these recipients through multiple mechanisms, often using more than one mechanism per initiative. As shown in figure 3, the initiatives primarily used grants and contracts. Of the 65 initiatives, 27 awarded grants, and 36 awarded contracts;[13] many awarded both. Agency officials also reported funding solar projects via cooperative agreements,[14] loans, and other mechanisms.[15]

[13]"Contract" means a mutually binding legal relationship obligating the seller to furnish the supplies or services (including construction) and the buyer to pay for them. It includes all types of commitments that obligate the Government to an expenditure of appropriated funds and that, except as otherwise authorized, are in writing. In addition to bilateral instruments, contracts include (but are not limited to) awards and notices of awards; job orders or task letters issued under basic ordering agreements; letter contracts; orders, such as purchase orders, under which the contract becomes effective by written acceptance or performance; and bilateral contract modifications. Contracts do not include grants and cooperative agreements covered by 31 U.S.C. 6301, et seq.

[14]Like grants, cooperative agreements involve the provision of financial or other support to accomplish a public purpose of support or stimulation authorized by federal statute. However, cooperative agreements differ from grants in terms of agency involvement, supervision, and intervention in the project. Whereas grants restrict government involvement to the minimum necessary to achieve program objectives, under cooperative agreements, the government and prime recipients share responsibility for the management, control, direction, and performance of projects.

[15]Officials from eight initiatives at three agencies reported that their initiative supported "other" funding mechanisms including administrative costs, cash prizes to competition winners, and laboratory work authorizations.

Figure 3: Number of Agencies' Solar-Related Initiatives Employing Various Funding Mechanisms

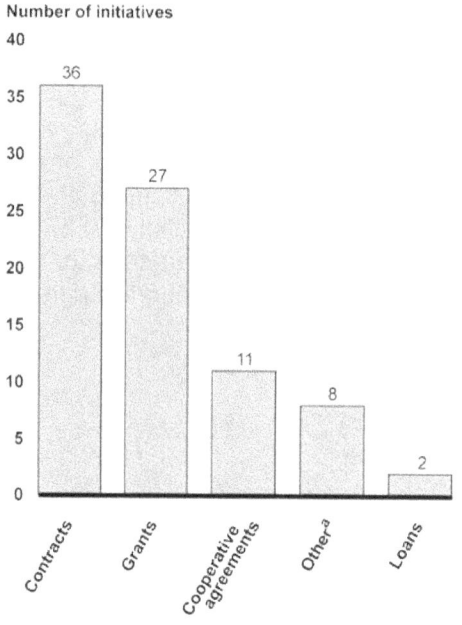

Number of initiatives

Funding mechanisms

Source: GAO analysis of agency-provided data

Note: Agency initiatives often used more than one funding mechanism, therefore the sum of initiatives across the agencies in this figure will not total 65, the number of initiatives in our review.

[a]Officials from eight initiatives at three agencies reported that their initiative supported "other" funding mechanisms including administrative costs, cash prizes to competition winners, and laboratory work authorizations.

Agency officials reported that the 65 initiatives as a group used multiple funding mechanisms, but we found that individual agencies tended to use primarily one or two funding mechanisms. For example, USDA exclusively used grants, while DOD tended to use contracts. DOE reported using grants and cooperative agreements almost equally. For example, DOE's *Solar ADEPT* initiative, an acronym for "Solar Agile Delivery of Electrical Power Technology," awards cooperative agreements to universities, industry, nonprofit organizations, and federal laboratories and researchers. Through a cooperative agreement, the initiative supported a project at the University of Colorado at Boulder that is developing advanced power conversion components that can be integrated into individual solar panels to improve energy yields. According to the project description, the power conversion devices will be designed for use on any type of solar panel. The University of Colorado at Boulder is partnering with industry and DOE's National Renewable Energy Laboratory on this project.

In responding to our questionnaire, officials from the six agencies reported that they obligated around $2.6 billion for the 1,506 solar projects in fiscal years 2010 and 2011. These obligations data represented a mix of actual obligations and estimates. Actual obligations were provided for both years for 51 of 65 initiatives. Officials provided estimated obligations for 12 initiatives for at least 1 of the 2 years, and officials from another 2 initiatives were unable to provide any obligations data.[16] Those officials who provided estimates or were unable to provide obligations data noted that the accuracy or the availability of the obligations data was limited because isolating the solar activities from the overall initiative obligations can be difficult. (See app. II for a full list of the initiatives and their related obligations.) As shown in table 2, over 90 percent of the funds (about $2.3 billion of $2.6 billion) were obligated by DOE. The majority of DOE's obligations (approximately $1.7 billion) were obligated as credit subsidy costs—the government's estimated net long-term cost, in present value terms, of the loans—as part of Title XVII Section 1705 Loan Guarantee Program from funds appropriated by Congress under the American Recovery and Reinvestment Act (Recovery Act).[17] Even excluding the Loan Guarantee Program funds, DOE obligated $661 million, which is more than was obligated by the other five agencies combined.

---

[16]Agency officials were asked to provide obligations data for fiscal years 2010 and 2011. In some cases, they provided estimated obligations for both years. In other cases, the officials were able to provide actual obligations data for 1 year and estimated obligations for the second. As a result, estimates were provided for 10 initiatives in fiscal year 2010 and for 9 initiatives in 2011. These estimates were for 12 initiatives.

[17]DOE's Loan Guarantee Program supports both commercialization and deployment technology advancement activities. The obligations provided by the agency and represented here do not distinguish between the commercialization and deployment technology advancement activities. In February 2009, the Recovery Act amended the Energy Policy Act of 2005, authorizing the Loan Guarantee Program to guarantee loans under section 1705. Funding was provided to pay credit subsidy costs for section 1705 projects—including those that support solar technologies—that began construction by September 30, 2011, and met other requirements. DOE estimated that the $2.5 billion in funding would be sufficient to provide about $18 billion in guarantees under section 1705. Section 1705 authorized guarantees for commercial energy projects that employ renewable energy systems, electric power transmission systems, or leading-edge biofuels that meet certain criteria. Credit subsidy costs exclude administrative costs and any incidental effects on governmental receipts or outlays. Present value is the worth of the future stream of returns or costs in terms of money paid immediately. In calculating present value, prevailing interest rates provide the basis for converting future amounts into their "money now" equivalents.

Table 2: Number of Federal Initiatives That Supported Solar Energy Technology for Fiscal Years 2010 and 2011, by Agency and Total Actual and Estimated Obligations

Agency	Number of initiatives for which data were provided	Actual obligations	Estimated obligations	Total obligations
DOE	18	$2,338,765,827[a]	$4,000,000	$2,342,765,827[b]
DOD	27	131,923,296	22,714,000	154,637,296
NSF	7	47,485,645	1,275,911	48,761,556
NASA	7	34,412,352	405,993	34,818,345
USDA	3	577,247	0	577,247
EPA	1	200,000	0	200,000
Total	63	$2,553,364,367	$28,395,904	$2,581,760,271

Source: GAO analysis of agency-provided data.

[a]Of the more than $2.3 billion obligated by DOE in 2010 and 2011 for solar projects, about $1.7 billion was obligated for credit subsidy costs for loan guarantees.

[b]Does not include two agency initiatives for which DOE did not provide obligations data.

Initiatives Are Fragmented across Agencies and Sometimes Overlap, but Agency Officials Reported Coordination to Avoid Duplication

The 65 solar-related initiatives are fragmented across six agencies and many overlap to some degree, but agency officials reported a number of coordination activities to avoid duplication. We found that many initiatives overlapped in the key characteristics of technology advancement activities, types of technologies, types of funding recipients, or broad goals; however, these areas of overlap do not necessarily lead to duplication of efforts because the initiatives sometimes differ in meaningful ways or leverage the efforts of other initiatives, and we did not find clear evidence of duplication among initiatives. Officials from most initiatives reported that they engage in a variety of coordination activities with other solar-related initiatives, at times specifically to avoid duplication.

Fragmented Initiatives Overlapped in Their Technology Advancement Activities, Types of Technologies, Funding Recipients, and Goals

The 65 solar-related initiatives are fragmented in that they are implemented by various offices across six agencies and address the same broad area of national need. In March 2011, we reported that fragmentation has the potential to result in duplication of resources.[18] However, such fragmentation is, by itself, not an indication that unnecessary duplication of efforts or activities exists. For example, in our March 2011 report, we stated that there can be advantages to having multiple federal agencies involved in a broad area of national need—

---

[18]GAO-11-318SP.

agencies can tailor initiatives to suit their specific missions and needs, among other things. In particular, DOD is able to focus its efforts on solar energy technologies that serve its energy security mission, among other things, and NASA is able to focus its efforts on solar energy technologies that aid in aeronautics and space exploration, among other things.

As table 3 illustrates, we found that many initiatives overlap because they support similar technology advancement activities and types of funding recipients. For example, initiatives that support basic and applied research most often fund universities, and those initiatives that support demonstration and commercialization activities most often fund industry.

Table 3: Number of Initiatives Supporting Technology Advancement Activities and Eligible Funding Recipients

Eligible funding recipient	Technology advancement activities			
	Basic research	Applied research	Demonstration activities	Commercialization
Federal laboratories/researchers	12	27	18	9
Industry	14	33	32	15
Nonprofit organizations	2	4	1	2
Universities	23	36	27	11
Other[a]	1	1	2	1

Source: GAO analysis of agency-provided data.

Note: Many solar-related initiatives support multiple technology advancement activities and multiple eligible funding recipients. Therefore, the totals in table 3 will not total 65, the number of initiatives in our review.

[a]Two initiatives supported other types of recipients: EPA's *P3: People, Prosperity & the Planet Student Design Competition for Sustainability* initiative supports teams of college students who participate in a competitive grants program in which they propose the research and development of sustainable designs related to water, energy, agriculture, the built environment, or materials and chemicals. DOD's *Smart Power Infrastructure Demonstration for Energy Reliability and Security (SPIDERS) Joint Capability Technology Demonstration* supports military installations to integrate renewable energy and storage in order to reduce the risk of extended electric grid outages.

Almost all of the initiatives overlapped to some degree with at least one other initiative in that they support broadly similar technology advancement activities, types of technologies, and eligible funding recipients.

- Twenty-seven initiatives support applied research for photovoltaic technologies by universities. For example, NSF's *Engineering Research Center for Quantum Energy and Sustainable Solar Technologies* at Arizona State University pursues cost-competitive photovoltaic technologies with sustained market growth. The Air Force's *Space Propulsion and Power Generation Research* initiative partners with various universities to develop improved methods for powering spacecraft, including solar cell technologies.

- Sixteen initiatives support demonstration activities focused on photovoltaic technologies by federal laboratories and researchers. For example, NASA's *High-Efficiency Space Power Systems* initiative conducts activities at NASA's Glenn Research Center to develop technologies to provide low cost and abundant power for deep space missions, such as highly reliable solar arrays, to enable a crewed mission to explore a near Earth asteroid. DOE's *Solar Energy Technologies Program (SETP),* which includes the *Photovoltaic Research and Development* initiative, works with national laboratories such as the National Renewable Energy Laboratory, Sandia National Laboratories, Brookhaven National Laboratory, and Oak Ridge Laboratory to advance a variety of photovoltaic technologies to enable solar energy to be as cost competitive as traditional energy sources by 2015.[19]

- Seven initiatives supported applied research on concentrated solar power technologies by industry. For example, DOE's *SETP Concentrated Solar Power* subprogram, which focuses on reducing the cost of and increasing the use of solar power in the United States, funded a company to develop the hard coat on reflective mirrors that is now being used in concentrated solar power applications. In addition, DOD's *Fast Access Spacecraft Testbed Program*, which concluded in March 2011, funded industry to demonstrate a suite of critical technologies including high-efficiency solar cells, sunlight concentrating arrays, large deployable structures, and ultra-lightweight solar arrays.

Additionally, 40 of the 65 initiatives overlap with at least one other initiative in that they supported similar broad goals, types of technologies, and technology advancement activities.

---

[19]In GAO-12-260, we separated SETP into four initiatives: Concentrated Solar Power, Photovoltaic Research and Development, Market Transformation, and Systems Integration.

- *Providing lightweight, portable energy sources.* Officials from several initiatives within DOD reported that their initiatives supported demonstration activities with the broad goal of providing lightweight, portable energy sources for military applications. For example, the goal of the Department of the Army's *Basic Solar Power Generation Research* initiative is to determine the feasibility and applicability of lightweight flexible, foldable solar panels for remote site power generation in tactical battlefield applications. Similarly, the goal of the Office of the Secretary of Defense's *Engineered Bio-Molecular Nano-Devices and Systems* initiative is to provide a low-cost, lightweight, portable photovoltaic device to reduce the footprint and logistical burden on the warfighter.

- *Artificial photosynthesis.* Several initiatives at DOE and NSF reported having the broad goal of supporting artificial photosynthesis, which converts sunlight, carbon dioxide, and water into a fuel, such as hydrogen. For example, one of DOE's *Energy Innovation Hubs, the Fuels from Sunlight Hub,* supports basic research to develop an artificial photosynthesis system with the specific goals of (1) understanding and designing catalytic complexes or solids that generate chemical fuel from carbon dioxide and/or water; (2) integrating all essential elements, from light capture to fuel formation components, into an effective system; and (3) providing a pragmatic evaluation of the system under development. NSF's *Catalysis and Biocatalysis* initiative has a specific goal of developing new materials that will be catalysts for converting sunlight into usable energy for direct use, or for conversion into electricity, or into fuel for use in fuel cell applications.

- *Integrating solar energy into the grid.* Officials from several initiatives reported focusing on demonstration activities for technologies with the broad goal of integrating solar or renewable energies into the grid or onto military bases. For example, DOE's *Smart Grid Research and Development* initiative has a goal of developing smart grid technologies,[20] particularly those that help match supply and demand in real time, to enable the integration of renewable energies, including solar energy, into the grid by helping stabilize variability and facilitate

---

[20]According to DOE, "smart grid" generally refers to a class of technology to bring utility electricity delivery systems into the twenty-first century, using computer-based remote control and automation. These systems are made possible by two-way communication technology and computer processing that has been used for decades in other industries.

GAO-12-843 Solar Energy

the safe and cost-effective operation by utilities and consumers. The goal of this initiative is to achieve a 20 percent improvement in the ratio of the average power supplied to the maximum demand for power during a specified period by 2020. DOD's *Installation Energy Research* initiative has a goal of developing better ways to integrate solar energy into a grid system, thereby optimizing the benefit of renewable energy sources.

Some initiatives may overlap on key characteristics such as technology advancement activities, types of technologies, types of recipients, or broad goals, but they also differ in meaningful ways that could result in specific and complementary research efforts, which may not be apparent when analyzing the characteristics. For example, an Army official told us that both the Army and Marine Corps were interested in developing a flexible solar substrate, which is a photovoltaic panel laminated onto fabric that can be rolled up and carried in a backpack. The Army developed technology that included a battery through its initiative, while the Marine Corps, through a separate initiative, altered the Army's technology to create a flexible solar substrate without a battery. Other initiatives may also overlap on key characteristics, but the efforts undertaken by their respective projects may complement each other rather than result in duplication. For example, DOE officials told us that one solar company may receive funding from multiple federal initiatives for different components of a larger project, thus simultaneously supporting a common goal without providing duplicative support.

While we did not find clear instances of duplicative initiatives, it is possible that there are duplicative activities among the initiatives that could be consolidated or resolved through enhanced coordination across agencies and at the initiative level. Also, it is possible that there are instances in which recipients receive funding from more than one federal source or that initiatives may fund some activities that would have otherwise sought and received private funding. Because it was beyond the scope of this work to look at the vast number of activities and individual awards that are encompassed in the initiatives we evaluated, we were unable to rule out the existence of any such duplication of activities or funding.

To Avoid Duplication, Agencies Coordinate and Verify That Recipients Did Not Receive Duplicative Funding

Officials from 57 of the 65 initiatives (88 percent) reported coordinating with other solar-related initiatives.[21] Coordination is important because, as we have previously reported, a lack of coordination can waste scarce funds and limit the overall effectiveness of the federal effort.[22] We have also previously reported that coordination across programs may help address fragmentation, overlap, and duplication.[23] Officials from nearly all initiatives that we identified as overlapping in their broad goals, types of technologies, and technology advancement activities, reported coordinating with other solar-related initiatives. In October 2005, we identified key practices that can help enhance and sustain federal agency coordination, such as (1) establishing joint strategies, which help align activities, core processes, and resources to accomplish a common outcome; (2) developing mechanisms to evaluate and report on the progress of achieving results, which allow agencies to identify areas for improvement; (3) leveraging resources, which helps obtain additional benefits that would not be available if agencies or offices were working separately; and (4) defining a common outcome, which helps overcome differences in missions, cultures, and established ways of doing business.[24] Agency officials at solar-related initiatives reported coordination activities that are consistent with these key practices, as described below.

---

[21]Officials were asked to report up to five examples of coordination activities with other solar-related initiatives. See app. III for questionnaire.

[22]GAO, *Results-Oriented Government: Practices That Can Help Enhance and Sustain Collaboration among Federal Agencies*, GAO-06-15 (Washington, D.C.: Oct. 21, 2005). The following additional key practices were noted in this report: (1) agreeing on roles and responsibilities, which clarifies who will do what, organizes joint and individual efforts, and facilitates decision making; (2) establishing compatible policies, procedures, and other means to operate across agency boundaries, which facilitates collaboration; (3) reinforcing agency accountability for collaborative efforts through agency plans and reports, to ensure that goals are consistent; and (4) reinforcing individual accountability for collaborative efforts through performance management systems, which strengthens accountability for results.

[23]GAO-11-318SP and GAO, *Employment For People With Disabilities: Little Is Known About the Effectiveness of Fragmented And Overlapping Programs*, GAO-12-677 (Washington, D.C.: June 29, 2012).

[24]GAO-06-15.

Some agency officials reported undertaking formal activities within their own agency to coordinate the efforts of multiple initiatives. For example:

- *Establishing a joint strategy.* NSF initiatives reported participating in an Energy Working Group, which includes initiatives in the agency's Directorates for Mathematical and Physical Sciences and for Engineering. Officials from initiatives we identified as overlapping reported participating in the Energy Working Group. NSF formed this group to initiate coordination of energy-related efforts between the two directorates, including solar efforts, and tasked it with establishing a uniform clean, sustainable energy strategy and implementation plan for the agency.

- *Developing mechanisms to monitor, evaluate, and report results.* DOD officials from initiatives in the Army, Marine Corps, and Navy that we identified as overlapping reported they participated in the agency's Energy and Power Community of Interest.[25] The goal of this group is to coordinate the R&D activities within DOD. The group is scheduled to meet every quarter, but an Army official told us the group has been meeting every 3 to 4 weeks recently to produce R&D road maps and to identify any gaps in energy and power R&D efforts that need to be addressed. Because of the information sharing that occurs during these meetings, the official said the risk of such duplication of efforts across initiatives within DOD is minimized.

In responding to our questionnaire, agency officials also reported engaging in formal activities across agencies to coordinate the efforts of multiple initiatives. For example:

- *Leveraging resources.* The Interagency Advanced Power Group (IAPG), which includes the Central Intelligence Agency, DOD, DOE, NASA, and the National Institute of Standards and Technology, is a federal membership organization that was established in the 1950s to streamline energy efforts across the government and to avoid duplicating research efforts. A number of smaller working groups were formed as part of this effort, including the Renewable Energy Conversion Working Group, which includes the coordination of solar

---

[25]Although it was not noted in the Air Force initiative's completed questionnaires, DOD officials told us the Air Force also participates in the Energy and Power Community of Interest.

efforts. The working groups are to meet at least once each year, but according to a DOD official, working group members often meet more often than that in conjunction with outside conferences and workshops. The purpose of the meetings is to present each agency's portfolio of research efforts and to inform and ultimately leverage resources across the participating agencies. According to IAPG documents, group activities allow agencies to identify and avoid duplication of efforts. Several of the initiatives that we identified as overlapping also reported participating in the IAPG.

- *Leveraging resources and defining a common outcome.* DOE's *SETP* in the Office of Energy Efficiency and Renewable Energy (EERE) coordinates with DOE's Office of Science and the Advanced Research Projects Agency-Energy (ARPA-E) through the SunShot Initiative, which according to SunShot officials, was established expressly to prevent duplication of efforts while maximizing agencywide impact on solar energy technologies. The goal of the SunShot Initiative is to reduce the total installed cost of solar energy systems by 75 percent. SunShot officials said program managers from all three offices participate on the SunShot management team, which holds "brain-storming" meetings to discuss ideas for upcoming funding announcements and subsequently vote on proposed funding announcements. Officials from other DOE offices and other federal agencies are invited to participate, with coordination occurring as funding opportunities arise in order to leverage resources. Officials said meetings may include as few as 25 or as many as 85 attendees, depending on the type of project and the expertise required of the attending officials. Additionally, DOE and NSF coordinate through the SunShot Initiative on the Foundational Program to Advance Cell Efficiency (F-PACE), which identifies and funds solar device physics and photovoltaic technology research and development that will improve photovoltaic cell performance and reduce module cost for grid-scale commercial applications. The initiatives that reported participating in SunShot activities also included many that we found to be overlapping.

- *Developing joint strategies; developing mechanisms to monitor, evaluate, and report results; and defining a common outcome.* The National Nanotechnology Initiative (NNI) an interagency program, which includes DOD, DOE, NASA, NSF, and USDA, among others, was established to coordinate the nanotechnology-related activities across federal agencies that fund nanoscale research or have a stake in the outcome of this research. The NNI is directed to (1) establish goals, priorities, and metrics for evaluation for federal nanotechnology research, development, and other activities; (2) invest in federal R&D programs in nanotechnology and related sciences to achieve these goals; and (3) provide for interagency coordination of federal nanotechnology research, development, and other activities. The NNI implementation plan states that the NNI will maximize the federal investment in nanotechnology and avoid unnecessary duplication of efforts. NNI includes a subgroup that focuses on nanotechnology for solar energy collection and conversion. Specifically, this subgroup is to (1) improve photovoltaic solar electricity generation with nanotechnology, (2) improve solar thermal energy generation and conversion with nanotechnology, and (3) improve solar-to-fuel conversions with nanotechnology.

In addition to the coordination efforts above, officials reported through our questionnaire that their agencies coordinate through discussions with other agency officials or as part of the program and project management and review processes. Some officials said such discussions and reviews among officials occur explicitly to determine whether there is duplication of funding occurring. For example, *SETP* projects include technical merit reviews, which include peer reviewers from outside of the federal government, as well as a federal review panel composed of officials from several agencies. Officials from *SETP* also participate in the technical merit reviews of other DOE offices' projects. ARPA-E initiatives also go through a review process that includes federal officials and independent experts. DOE officials told us that an ARPA-E *High Energy Advanced Thermal Storage* review meeting, an instance of potential duplicative funding was found with an *SETP* project. Funding of the project through *SETP* was subsequently removed because of the ARPA-E review process, and no duplicative funds were expended.

In addition to coordinating to avoid duplication, officials from 59 of the 65 initiatives (91 percent) reported that they determine whether applicants have received other sources of federal funding for the project for which they are applying. Twenty-one of the 65 initiatives (32 percent) further reported that they have policies that either prohibit or permit recipients from receiving other sources of federal funding for projects. Some respondents to our questionnaire said it is part of their project management process to follow up with funding recipients on a regular basis to determine whether they have subsequently received other sources of funding. For example, DOE's ARPA-E prohibits recipients from receiving duplicative funding from either public or private sources, and requires disclosure of other sources of funding both at the time of application, as well as on a quarterly basis throughout the performance of the award. Even if an agency requires that such funding information be disclosed on applications, applicants may choose not to disclose it. In fact, it was recently discovered that a university researcher did not identify other sources of funding on his federal applications as was required and accepted funding for the same research on solar conversion of carbon dioxide into hydrocarbons from both NSF and DOE. Ultimately, the professor was charged with and pleaded guilty to wire fraud, false statements, and money laundering in connection with the federal research grant.

Agency Comments and Our Evaluation

We provided DOD, DOE, EPA, NASA, NSF, and USDA with a draft of this report for review and comment. USDA generally agreed with the overall findings of the report. NASA and NSF provided technical or clarifying comments, which we incorporated as appropriate. DOD, DOE, and EPA indicated that they had no comments on the report.

As agreed with your offices, unless you publicly announce the contents of this report earlier, we plan no further distribution until 30 days from the report date. At that time, we will send copies to the Secretaries of Agriculture, Defense, and Energy; the Administrators of EPA and NASA; the Director of NSF; the appropriate congressional committees; and other interested parties. In addition, the report will be available at no charge on the GAO website at http://www.gao.gov.

If you or your staff members have any questions about this report, please contact me at (202) 512-3841 or ruscof@gao.gov. Contact points for our Offices of Congressional Relations and Public Affairs may be found on the last page of this report. GAO staff who made key contributions to this report are listed in appendix IV.

Frank Rusco
Director
Natural Resources and Environment

Appendix I: Scope and Methodology

The objectives of our report were to identify (1) solar-related initiatives supported by federal agencies in fiscal years 2010 and 2011 and key characteristics of those initiatives and (2) the extent of fragmentation, overlap, and duplication, if any, among federal solar-related initiatives, as well as the extent of coordination among these initiatives.

To inform our objectives, we reviewed a February 2012 GAO report that was conducted to identify federal agencies' renewable energy initiatives, which included solar-related initiatives, and examine the federal roles the agencies' initiatives support.[1] The GAO report on renewable energy-related initiatives identified nearly 700 initiatives[2] that were implemented in fiscal year 2010 across the federal government, of which 345 initiatives supported solar energy. For purposes of this report, we only included those solar-related initiatives that we determined were focused on research and development (R&D), and commercialization, which we defined as follows:

- *Research and development.* Efforts ranging from defining scientific concepts to those applying and demonstrating new and improved technologies.

- *Commercialization.* Efforts to bridge the gap between research and development activities and the marketplace by transitioning technologies to commercial applications.

We did not include those initiatives that focused solely on deployment activities, which include efforts to facilitate or achieve widespread use of existing technologies either in the commercial market or for nonmarket uses such as defense, through their construction, operation, or use. Initiatives that focus on deployment activities include a variety of tax incentives. We also narrowed our list to only those initiatives that focused research on advancing or developing new and innovative solar technologies.

---

[1] GAO-12-260.

[2] For purposes of GAO-12-260, and therefore this report as well, a renewable energy initiative is defined as a program or group of agency activities serving a similar purpose or function that involved renewable energy through a specific emphasis or focus, even if renewable energy was only one part of a broader effort.

Next, we shared our list with agency officials and provided our definitions of R&D and commercialization. We asked officials to determine whether the list was complete and accurate for fiscal year 2010 initiatives that met our criteria, whether those initiatives were still active in fiscal year 2011, and whether there were any new initiatives in fiscal year 2011. If officials wanted to remove an initiative from our list, we asked for additional information to support the removal. In total, we determined that there were 65 initiatives that met our criteria.

To identify and describe the key characteristics of solar-related initiatives implemented by federal agencies, we developed a questionnaire to collect information from officials of those 65 federal solar energy-related initiatives. The questionnaire was prepopulated with information that was obtained from the agencies for GAO's renewable energy report including program descriptions, type of solar technology supported, funding mechanisms, and type of funding recipients. Questions included the type of technology advancement activities, obligations for solar activities in fiscal years 2010 and 2011, initiative-wide and solar-specific goals, and coordination efforts with other solar-related initiatives. We conducted pretests with officials of three different initiatives at three different agencies to check that (1) the questions were clear and unambiguous, (2) terminology was used correctly, (3) the questionnaire did not place an undue burden on agency officials, (4) the information could feasibly be obtained, and (5) the questionnaire was comprehensive and unbiased. An independent GAO reviewer also reviewed a draft of the questionnaire prior to its administration. On the basis of feedback from these pretests and independent review, we revised the survey in order to improve its clarity.

After completing the pretests, we administered the questionnaire. We sent questionnaires to the appropriate agency liaisons in an attached Microsoft Word form, who in turn sent the questionnaires to the appropriate officials. We received questionnaire responses for each initiative and, thus, had a response rate of 100 percent. After reviewing the responses, we conducted follow-up e-mail exchanges or telephone discussions with agency officials when responses were unclear or conflicting. When necessary, we used the clarifying information provided by agency officials to update answers to questions to improve the accuracy and completeness of the data.

Because this effort was not a sample survey, it has no sampling errors. However, the practical difficulties of conducting any survey may introduce errors, commonly referred to as nonsampling errors. For example, difficulties in interpreting a particular question, sources of information available to respondents, or entering data into a database or analyzing them can introduce unwanted variability into the survey results. However, we took steps to minimize such nonsampling errors in developing the questionnaire—including using a social science survey specialist for design and pretesting the questionnaire. We also minimized the nonsampling errors when collecting and analyzing the data, including using a computer program for analysis, and using an independent analyst to review the computer program. Finally, we verified the accuracy of a small sample of keypunched records by comparing them with their corresponding questionnaires, and we corrected the errors we found. Less than 0.5 percent of the data items we checked had random keypunch errors that would not have been corrected during data processing. To conduct our analysis, a technologist compared all of the initiatives and identified overlapping initiatives as those sharing at least one common technology advancement activity, one common technology, and having similar goals. A second technologist then completed the same analysis, and the two then compared their findings and, where they differed, came to a joint decision as to which initiatives broadly overlapped on their technology advancement activities, technologies, and broad goals. If the two technologists could not come to an agreement, a third technologist determined whether there was overlap. To assess the reliability of obligations data, we asked officials of initiatives that comprised over 90 percent of the total obligations follow-up questions on the data systems used to generate that data. While we did not verify all responses, on the basis of our application of recognized survey design practices and follow-up procedures, we determined that the data used in this report were of sufficient quality for our purposes.

We conducted this performance audit from September 2011 to August 2012 in accordance with generally accepted government auditing standards. Those standards require that we plan and perform the audit to obtain sufficient, appropriate evidence to provide a reasonable basis for our findings and conclusions based on our audit objectives. We believe that the evidence obtained provides a reasonable basis for our findings and conclusions based on our audit objectives.

Appendix II: Solar-Related Initiatives at Six Federal Agencies, Descriptions and Fiscal Years 2010 and 2011 Obligations

Tables 4, 5, 6, 7, 8, and 9 provide descriptions, by agency, of the 65 initiatives that support solar energy technologies and the obligations for those initiatives' solar activities in fiscal years 2010 and 2011.

Table 4: Department of Defense Solar-Related Initiatives

Initiative name and implementing organization	Description	FY 2010 and FY 2011 total (actual and estimated) obligations
Air Force		
Aviation Propulsion / Power Generation Research	Renewable fuels evaluation is conducted as part of component testing for new propulsion and power systems. The initiative's goals are to carry out research, development, and demonstration efforts to integrate high-efficiency, lightweight, and flexible solar cells into operational small unmanned air systems in order to extend flight endurance. The Air Force also conducts congressional or committee-directed work.	$7,086,000
Other Power Generation Research	This initiative comprises several basic research efforts related to improving materials to be used in producing renewable energy. Renewable energy-related research under this program includes, for example, efforts to design organic solar cells and to design new nanostructures for conversion of solar energy to electricity. Solar-related research includes developing the capability to produce low-cost, lightweight, flexible, and efficient solar cells with numerous military applications including powering of autonomous unmanned air vehicles, remote sensors, and deployed systems. A variety of next-generation, advanced solar cell technologies are being pursued including both polymer-based and nanoparticle-based devices.	8,100,000[a]
Other Research and Development Efforts	This area includes several Air Force research efforts, most of which are congressional or committee-directed, that involve renewable energy. For example, as part of the Support Systems Development Program's Logistics Systems Development Project, Air Force efforts include alternate energy research and integration, and demonstration and validation of renewable energy technology. Specific projects will validate that on-base or near-base deployment of thin-film solar photovoltaic arrays and concentrated solar designs are a cost effective solution that will allow the Air Force installations at-large to achieve renewable energy mandates, while also providing a secure clean and reliable back-up source of energy.	6,290,643
Power Generation Research for Facilities and Deployed Locations	This area of Air Force research includes activities to reduce costs and improve the performance and sustainability of Air Force operations. Renewable energy-related research in this area includes specific efforts to develop higher efficiency solar power technologies for deployed applications. Also, as part of programs under this area, the Air Force is charged with administering some congressional or committee-directed work that, in some instances, includes a renewable energy-related focus. The goal for the Solar Integrated Power Shelter System is to generate 3kW power and reduce demands by 50% to allow cooling two shelters with a single environmental control unit. In this project, deployable shelters are retrofit with improved or emerging shelter liner (insulation), shelter fly, and/or thin photovoltaic modules and evaluated for their effectiveness to improve efficiency and reduce energy consumption.	2,270,000[a]

Initiative name and implementing organization	Description	FY 2010 and FY 2011 total (actual and estimated) obligations
Space Propulsion / Power Generation Research	This area of Air Force research includes several programs with work that includes developing improved methods for powering spacecraft. Renewable energy-related research under this area is focused on the use of solar power with the goal of developing solar cells that are higher in efficiency and lower in mass while providing flexibility for integration into novel array structures. For example, under the Space Technology and Advanced Spacecraft Technology Programs, the Air Force is working to develop and improve solar cell technologies for space applications. Also, as part of this area of research, the Air Force is charged with administering some congressional or committee-directed work that, in some instances, includes a renewable energy-related focus.	17,517,562
Army		
Basic Solar Power Generation Research	Army research efforts in this area involve two programs that fund applied research on technologies in areas such as electronic and power components, as well as demonstration of more advanced military engineering technologies. Renewable energy-related research activities in this area include investigating the use of solar energy technologies for various Army applications. These efforts are funded as part of congressional or committee-directed work that the Army is charged with administering. For example, in fiscal year 2010, the Army funded research on advanced flexible solar photovoltaic technologies, among other projects related to solar energy.	5,770,928
Installation Energy Research	This area of Army research includes programs that develop and demonstrate military engineering technologies, among other efforts. Renewable energy-related research activities in this area involve developing energy technology solutions, including renewable energy, for fixed installations and enduring forward operating bases. For example, as part of the Military Engineering Technology Program, the Army researches technologies necessary for secure, energy efficient, sustainable military installations, including developing methods to optimize sustainable energy generation and integrate renewable energy resources. Also, as part of these programs, the Army is charged with administering some congressional or committed-directed work that, in some instances, includes a renewable energy-related focus. Specific solar goals for this initiative include developing better ways to integrate solar energy into a grid system and improving the conversion efficiency of solar photovoltaic devices for creating novel photovoltaic configurations allowing flexible mounting of solar cells onto Army systems.	850,500
Intelligent Power Distribution System Research	Army research efforts in this area span several programs that focus on technology development and demonstration to meet a variety of military needs. Renewable energy-related research activities in this area focus on developing power distribution systems that can accept inputs from a variety of energy sources, including alternative/renewable energy sources. For example, as part of the Army's work on the Transportable Hybrid Electric Power Station project, the Army is demonstrating intelligent power management technology that helps incorporate renewable energy sources to reduce the use of fossil fuel generators. Also, as part of these programs, the Army is charged with administering some congressional or committed-directed work that, in some instances, includes a renewable energy-related focus. For example, in fiscal year 2010, this work included an effort to build and demonstrate a microgrid with plug-in electric vehicles to determine the technical readiness of microgrids to accept power from various inputs while charging the selected vehicles and providing power to other applications.	1,372,000[a]

Initiative name and implementing organization	Description	FY 2010 and FY 2011 total (actual and estimated) obligations
National Defense Center for Energy and Environment	This Army research effort includes funding for the National Defense Center for Energy and Environment, which was established in 1990 to address high-priority environmental problems for DOD, other government agencies, and the industrial community. The center is used to demonstrate and validate environmentally acceptable technologies and assist in the training of potential users as part of the technology transfer process. The center is managed by the Army on behalf of the Office of the Deputy Under Secretary of Defense for Installations and Environment. The center conducts work at the request of multiple stakeholders, including, among other topics, research related to alternative fuels and renewable energy. While not specifically focused or directly related to solar energy research and/or technology, the goal of this program in terms of solar energy is to demonstrate renewable energy solutions (that include solar technology) and capture technology performance and cost-related data that can be used by the customer as a basis for determining if the technology is viable for their installation or operational needs.	1,775,000[a]
Other Power Generation Research	Army research efforts in this area encompass two programs; one that focuses on fundamental research to provide new concepts and technologies to meet the Army's future needs, and another that supports advanced development of technologies to reduce the logistics burden. Renewable energy-related research activities in this area focus on using renewable energy sources for power generation. However, unlike Army renewable energy-related research in other areas, research in this area examines renewable energy sources in general and is not specifically targeted to certain renewable energy sources. For example, under the Defense Research Sciences Program, the Army funds the investigation of a variety of approaches for novel energy harvesting (e.g., from light, heat, vibration, isotope, and biological energy sources). Alternatively, under the Soldier Support and Survivability Program, the Army has funded demonstrations of improvements to combat feeding equipment to improve the energy efficiency of refrigeration equipment to allow the equipment to be powered by renewable energy sources (e.g., solar power). Current solar-related activities include efforts to improve the energy conversion efficiency of photovoltaic devices to create novel photovoltaic configurations allowing flexible and conformal mounting of solar cells onto Army systems. Thermophotovoltaic energy sources are also being examined.	389,000
Other Vehicle Energy Storage Research	Army research in this area involves the Combat Vehicle and Automotive Technology Program, which funds development of automotive technologies through various efforts, including, for example, the National Automotive Center—a shared government and industry program to leverage commercial investments in automotive technology research for Army ground combat and tactical vehicle applications. Renewable energy-related research in this area focuses on advancing vehicle energy storage technologies that will allow vehicles to incorporate alternative/renewable energy into hybrid electric vehicles.	3,526,831

Initiative name and implementing organization	Description	FY 2010 and FY 2011 total (actual and estimated) obligations
Tactical Electric Power Research	Army efforts in this area span several research programs that develop, mature, and accelerate technologies to meet a variety of military needs. Renewable energy-related research activities in this area focus on incorporating renewable energy technologies into tactical electric power generation. For example, the Combating Terrorism-Technology Development Program includes an effort to develop a Transportable Hybrid Electric Power Station that incorporates solar and wind energy technologies to reduce use of fossil fuel generators. As part of these programs, the Army is charged with administering some congressional or committee-directed work that, in some instances, includes a renewable energy-related focus. For example, in fiscal year 2010, this work included developing a flexible solar cell for a man portable power generator.	3,100,000
Marine Corps		
Enhanced Company Operations Small Unit Power and Logistics Demand Reduction	This initiative represents a series of Marine Corps efforts to examine the training, tactics, equipment, and other aspects of the operations of a Marine Corps infantry company in order to identify ways to improve the ability of the company to operate independently. In fiscal year 2010, these efforts included assessment of renewable energy options to include the use of a reformed methanol fuel cell to support a small unit electrical power system that could reduce the number, weight, and types of replacement batteries carried by the Marine Corps. Another renewable energy-related effort under this initiative involves examining ways to reduce the logistics demand of a company-sized unit in a variety of ways, including the use of renewable energy technologies. Solar energy is one of several areas explored for the Logistics Demand Reduction initiative. All of these efforts support the goal identified in the Marine Corps Expeditionary Energy Strategy of meeting operational demand with renewable energy.	18,790,000
Experimental Forward Operating Base	This initiative is a Marine Corps process that brings together stakeholders from across the Marine Corps' requirements, acquisitions, and technology development communities to quickly evaluate and deploy technologies to reduce the need for battlefield "liquid logistics" (fuel and water) and to inform requirements development. For example, in August 2010, a Marine unit tested three renewable technologies— the Ground Renewable Expeditionary Energy System, the Solar Portable Alternative Communications Energy System, and the Zerobase Solar Regenerator.	1,886,000
Navy		
Future Naval Capabilities Program	The Future Naval Capabilities Program represents the requirements-driven, delivery oriented portion of the Navy's Science and Technology portfolio. Future Naval Capabilities investments respond to Naval Science and Technology Gaps that are generated by the Navy and Marine Corps after receiving input from Naval Research Enterprise stakeholders. The Future Naval Capabilities Program focuses on high-priority capabilities and transitioning developed products to naval acquisition and naval forces. This program focuses on taking technologies that have reached the applied or advanced technology development research stages, and maturing them for transition to Naval acquisition programs. Two areas of focus under the program include (1) "Lightening the Load" and (2) "Battlefield Power." In fiscal years 2010 and 2011, the Advanced Power Generation product did have one solar-related initiative: the Ground Renewable Expeditionary Energy System (GREENS). This portable hybrid photovoltaic/battery power system was developed and demonstrated for the Marine Corps and was subsequently deployed forward as part of Operation Enduring Freedom. The solar research goal under this initiative was to develop and demonstrate a photovoltaic energy system capable of providing 300 watts of power continuously for 24 hours per day/7 days per week and suitable for fielding by the Marine Corps.	700,000[a]

Initiative name and implementing organization	Description	FY 2010 and FY 2011 total (actual and estimated) obligations
Other Power Generation Research	This initiative has a goal of discovering and developing new power generation materials, devices and systems, maturing power generation concepts capable of meeting application specific metrics and constraints to Technology Readiness Level (TRL) 6, and transitioning those new energy storage technology capabilities to Navy and Marine Corps acquisition program offices for final maturation and fielding (TRL 7-9). The goal for this solar initiative is to develop low cost, lightweight, rollable, and easily deployable solar cells that can greatly reduce the logistics burden of supplying power to the Marine Corps and could also power remote sensors for the Navy. These goals include the development of new active materials by computation, organic synthesis, and characterization, as well as materials and technology scale-up and development of devices with improved stability.	6,825,000[a]
Office of the Secretary of Defense		
Engineered Bio-Molecular Nano-Devices and Systems	This initiative is part of the Materials Processing Technology project under Alternate Power Sources which includes the Defense Advanced Research Projects Agency (DARPA) Portable Photovoltaic Program. Other programs in the Alternate Power Sources heading are not solar-related. The overall goal of DARPA's Portable Photovoltaic Program is to provide a low-cost, lightweight, portable photovoltaic device that has high-power conversion efficiency in a form factor amenable to low-cost production on flexible substrates for widespread use by DOD forces to reduce the footprint and logistical burden on the warfighter.	3,256,201
Environmental Security Technology Certification Program – Energy and Water Program	This initiative is a DOD demonstration/validation program, under which DOD provides funding to take environmental and energy technologies that are ready to be demonstrated and places them at DOD facilities for the purpose of collecting data on cost and performance to see if the technologies should be brought to scale and used. Demonstration projects are selected from across the federal government and the private sector using a competitive solicitation process. There are five program areas under the initiative, including an Energy and Water Program, which provides funding for demonstrations of renewable energy technologies, among other types of energy and water projects. The demonstrations help increase the speed and scale of commercialization of emerging renewable energy technologies by helping to build acceptance of the technologies through partnerships between technology developers and DOD. Solar energy, including technologies such as solar thermal technologies for hot water and HVAC applications, is a subset of efforts to find cost competitive distributed energy generation that will improve DOD installation energy security.	10,400,000
Fast Access Spacecraft Testbed Program	The Fast Access Spacecraft Testbed (FAST) Program was part of the Space Programs and Technology project. Its goal was to demonstrate a suite of critical technologies, including high-efficiency solar cells, sunlight concentrating arrays, and ultra lightweight solar arrays. Combined with electric propulsion, these technologies would enable fast-transfer roaming satellites with nearly five times the fuel efficiency of conventional chemical propulsion.	3,060,000

Initiative name and implementing organization	Description	FY 2010 and FY 2011 total (actual and estimated) obligations
Mobile Smart Power Initiative	This initiative was started to demonstrate alternative power generation options for small forward operating bases. These efforts included assessing the performance, safety, reliability, and maintainability of a hybrid photovoltaic battery system. Special Operations Command evaluated several different systems as part of this initiative under two programs: one that provides funding to demonstrate and evaluate emerging/advanced technologies for Special Operations Forces; and another that provides funding to develop and deploy special capabilities for Special Operations Forces to perform intelligence surveillance and reconnaissance. Special Operations Command sought to demonstrate that larger scale solar power (28kW - 56kW) systems could save fuel and lives at forward operating bases in austere locations such as Afghanistan.	1,650,000
Net Zero Plus Joint Capability Technology Demonstration Program	This program has demonstrated various technologies to reduce fuel use at forward operating bases. It leverages research and development from federal and private labs and available technologies in, for example, alternative fuels, novel power storage, and innovative power generation. Technologies ranged from energy efficient structures to smart power management. The renewable energy technologies consisted of solar panels built into tent structures and 200 kW solar panel arrays integrated with the microgrid to reduce fuel usage. The program is part of the Joint Capability Technology Demonstration Program, which demonstrates joint solutions to prioritized combatant commander capability gaps and speeds solutions to warfighters in 18 to 36 months. The Joint Capability Technology Demonstration Program goal for solar energy technology included incorporation of flexible photovoltaic panels to utilize solar energy as a renewable energy source to provide electricity that would normally be supplied by conventional generators.	2,442,000
Novel Power Sources	This initiative is part of the Materials Processing Technology project, which seeks to develop materials that will lower the cost, increase the performance, or enable new missions for military platforms and systems. Under this initiative, DARPA explores new materials and processes to efficiently generate and control power, with a focus on energy sources that are compatible with military fuels. Renewable energy efforts under this initiative include developing and integrating new processes to convert carbon dioxide and water into syngas and cellulosic biomass into synthetic fuel for use in fuels cells, biomass conversion systems, and solar fuel systems.	16,511,706
Photovoltaic Power Supply	As part of the work under its Advanced Technology Development Program, Special Operations Command is charged with administering some congressional or committee-directed work. In some instances, this work includes a renewable energy-related focus. For example, in fiscal year 2010, this work included conducting research and development of technologies to improve the efficiency of solar photovoltaic panels for small scale charging systems. The goal of this project is to develop increased efficiency in photovoltaic cell technology for Special Operations Command applications, including developing various photovoltaic compositions to increase solar cell conversion efficiency and to assess their ability to power autonomous sensors.	1,970,851
Photovoltaic Ribbon Solar Cell Technology Project	As part of the work under its Materials Processing Technology project, the Defense Advanced Research Projects Agency is charged with administering some congressional or committee-directed work. In some instances, this work includes a renewable energy-related focus. For example, in fiscal year 2010, this work included conducting research into photovoltaic ribbon solar cell technology. The goal of this project is to create a new process for synthesis of thin, flexible, high-quality solar cells. The solar cells will be appropriate for coupling with prismatic holographic films for the purpose of creating lightweight, flexible modules.	2,869,963

Initiative name and implementing organization	Description	FY 2010 and FY 2011 total (actual and estimated) obligations
Smart Power Infrastructure Demonstration for Energy Reliability and Security (SPIDERS) Joint Capability Technology Demonstration	SPIDERS will demonstrate cyber-secure "smart" microgrids with demand side management and integration of renewable energy and storage on military installations, in partnership with Department of Homeland Security and Department of Energy. SPIDERS will reduce the "unacceptably high risk" of extended electric grid outages by developing the capability to "island" installations while maintaining operational surety and security. SPIDERS will protect task critical assets from loss of power due to cyber attack, integrate renewable (solar, wind) and other distributed energy generation concepts to power task critical assets in times of emergency, sustain critical operations during prolonged power outages, and manage installation electrical power and consumption efficiency to reduce petroleum demand and carbon "bootprint".	4,421,000
Very High Efficiency Solar Cell Program	This program addresses a variety of issues with improving the efficiency of photovoltaic technologies, including the development of high-efficiency design concepts; new and innovative components, materials, and processes necessary to achieve these concepts; and scalable fabrication processes that allow industrial manufacturing of an affordable product. In particular, work under this program seeks to raise the efficiency of a new class of solar modules to 40 percent and deliver engineering prototypes. The technology would be used to power both permanent and mobile bases, as well as to reduce the logistical burden of supplying energy (e.g., batteries and fuel) in the field.	4,753,111
Vulture Program	The objective of the Vulture program is to develop and demonstrate the technology to enable an airborne payload to remain persistently on-station, uninterrupted and unreplenished, for over 5 years performing strategic and tactical communications, position/navigation/timing and intelligence, surveillance, and reconnaissance missions over an area of interest. Vulture technology enables a retaskable, persistent pseudo-satellite capability, in an aircraft package. The program is pursuing technologies that advance the state of the art for a self-contained solar regenerative closed cycle energy system to be used to propel and provide prime power on long endurance air platforms. To this end, the program plans to integrate and test commercially available technologies that include solar and solid oxide fuel cells.	17,053,000
Total		$154,637,296

Source: GAO analysis of agency-provided data.

[a]All or part of the agency-provided obligations are estimated.

Table 5: Department of Energy Solar-Related Initiatives

Initiative name and implementing office	Description	FY 2010 and FY 2011 total (actual and estimated) obligations
Advanced Research Projects Agency-Energy		
ARPA-E Funding Opportunity Announcement 1[a]	Under this initiative, ARPA-E provides funding to support research on a variety of energy ideas and technologies. This research funding is focused on applicants with well-formed research and development plans for potentially high-impact concepts or new technologies, including renewable technologies.	$5,000,272
High Energy Advanced Thermal Storage (HEATS)[a]	Through this initiative, ARPA-E seeks to develop revolutionary cost-effective thermal energy storage technologies in three focus areas: (1) high temperature storage systems to deliver solar electricity more efficiently around the clock and allow nuclear and fossil baseload resources the flexibility to meet peak demands, (2) fuel produced from the sun's heat, and (3) HVAC systems that use thermal storage to dramatically improve the driving range of electric vehicles.	0
Solar Solar Agile Delivery of Electrical Power Technology (Solar ADEPT)[a]	Solar ADEPT is ARPA-E's portion of the SunShot collaboration. SunShot leverages the unique strengths across DOE to reduce the total cost of utility-scale solar systems by 75 percent by 2017. If successful, this collaboration would deliver solar electricity at roughly 6 cents a kW hour—a cost competitive with fossil fuels. Solar ADEPT focuses on integrating advanced power electronics into solar panels and solar farms to extract and deliver energy more efficiently. Specifically, ARPA-E aims to invest in key advances in magnetic, semiconductor switches, and charge storage, which could reduce power conversion costs by up to 50 percent for utilities and 80 percent for homeowners.	0
Energy Efficiency and Renewable Energy		
Emerging Technologies	Through this initiative, EERE works to develop cost-effective technologies to meet the thermal energy needs of residential and commercial buildings, including through solar heating and cooling. Specific activities related to renewable energy include funding federal researchers and national labs to analyze technology advancement of solar domestic water heating, space heating, and space cooling, and determine future needs for creating marketable net zero energy designs—designs which allow buildings to produce as much energy as they use.	5,195,000
Hydrogen Fuel R&D[b]	This initiative focuses on R&D to develop hydrogen production, delivery, and storage technologies, including those from hydrogen produced using renewable energy sources such as biomass, solar, wind, etc.	10,102,840
Hydrogen and Fuel Cell Technologies Crosscutting Activities[b]	This initiative focuses on crosscutting functions for hydrogen and fuel cell technologies such as manufacturing, technology validation, safety codes and standards, and education. It also analyzes technologies to identify gaps and help direct future research and development.	0
National Renewable Energy Laboratory's Laboratory Directed Research and Development	Projects through this initiative serve as the basis for proposing new projects and/or programs to DOE by developing and demonstrating a new capability or establishing proof-of-principle at the forefront of science and technology. Projects, which must adhere to the laboratory-directed research and development policy, are selected based on technical merit and strategic alignment and are screened to assure they are not duplicative of program efforts.	6,019,591

Initiative name and implementing office	Description	FY 2010 and FY 2011 total (actual and estimated) obligations
Solar Energy Technologies Program – Concentrated Solar Power	This initiative focuses on concentrating solar power technologies, which use mirrors to reflect and concentrate sunlight onto receivers that collect the solar energy and convert it to heat. Goals include increasing the use of these technologies in the United States, making them competitive in the intermediate power market—which operates between the continuously operating baseload market and the high demand only peak market—by 2015, and developing advanced technologies that will reduce systems and storage costs to make them competitive in the baseload power market by 2020. Efforts include contracts with industry, advanced research at its national laboratories, and collaboration with other government agencies to remove barriers to deploying the technology.	78,722,511
Solar Energy Technologies Program – Photovoltaic R&D	This initiative advances photovoltaic technology with a goal of achieving grid parity—the point at which alternative means of generating electricity are as cheap as traditional sources—by 2015. To that end, DOE is investing in approaches ranging from basic cell technologies to total system development that demonstrate progress toward minimizing the life cycle cost of solar energy. In conducting this work, DOE is partnering with its national laboratories, start-up companies, universities, and integrated industry teams.	260,829,432
Solar Energy Technologies Program – Market Transformation	This initiative identifies and addresses market barriers to make it possible for states, cities, and utilities to adopt solar energy programs. It develops specific activities and external partnerships to address those barriers—for example, opening the grid for electricity produced by solar power plants.	39,608,824
Solar Energy Technologies Program – Systems Integration	Under this initiative, DOE works with the solar industry, utilities, and national laboratories to address the barriers to large-scale deployment of solar technologies. It focuses on understanding and removing the regulatory, technical, and economic barriers to integrating solar electricity into the electric grid.	76,764,397
Loan Programs Office		
Title XVII Section 1703 Loan Guarantee Program	Section 1703 of Title XVII of the Energy Policy Act of 2005 authorized DOE to provide loans to support innovative clean energy technologies that are typically unable to obtain conventional private financing due to high technology risks. In addition, the technologies must avoid, reduce, or sequester air pollutants or emissions of greenhouse gases. Eligible technologies under this initiative included biomass, solar, wind, hydropower, and alternative fuel vehicles, among others. Under section 1703, borrowers were to pay DOE for the credit subsidy costs. However, DOE noted that, once the section 1705 Loan Guarantee Program was established under the Recovery Act specifically to support renewable energy and certain other projects, renewable energy projects have been supported under the section 1705 Loan Guarantee Program rather than the section 1703 Loan Guarantee Program.	0
Title XVII Section 1705 Loan Guarantee Program[c]	The Recovery Act added section 1705 to the Energy Policy Act of 2005. Section 1705 authorizes a temporary program to provide loan guarantees for certain renewable energy systems, electric power transmission systems and innovative biofuel projects that began construction no later than September 30, 2011. To implement this authority, DOE established the Financial Institution Partnership Program, a risk-sharing partnership between DOE and qualified finance organizations. Through this program, DOE pays the credit subsidy costs of loan guarantees using funds appropriated for this purpose and guarantees up to 80 percent of a loan provided for a renewable energy generation project by qualified financial institutions. In addition, DOE deployed the section 1705 authority to fund eligible innovative solar projects through the Federal Financing Bank.	1,681,764,960

Initiative name and implementing office	Description	FY 2010 and FY 2011 total (actual and estimated) obligations
Office of Electricity Delivery and Energy		
Energy Storage	This initiative focuses on storing electricity generated by renewable energy sources during periods of low demand and discharging it during periods of high demand. It also focuses on connecting renewable sources to the grid, allowing transmission lines to be more effectively utilized, and supporting demonstration projects. The Office of Electricity Delivery and Energy's efforts under this initiative also include communications, education, and outreach.	**4,000,000[a]**
Clean Energy Transmission and Reliability	This initiative focuses on developing tools and techniques to address challenges that renewable energy sources—particularly wind and solar power—pose to electricity grid operators, who need to incorporate these variable generation sources without compromising reliability. The Office of Electricity Delivery and Energy's efforts under this initiative also include communications, education, and outreach.	**Unable to provide**
Smart Grid Research and Development	This initiative focuses on developing smart grid technologies, particularly those that help match supply and demand in real time. Such technologies enable the integration of renewable resources by reducing power disturbances, helping stabilize the variability of renewable energy sources, and facilitating their safe and cost-effective operation by utilities and consumers. The Office of Electricity Delivery and Energy's efforts under this initiative also include communications, education, and outreach.	**Unable to provide**
Office of Science		
Office of Basic Energy Sciences - Chemical Sciences, Geosciences, and Biosciences Core Research	The Office of Basic Energy Sciences supports fundamental research to understand, predict, and ultimately control matter and energy at the electronic, atomic, and molecular levels in order to provide the foundations for new energy technologies and to support DOE missions in energy, environment, and national security. The Chemical Sciences, Geosciences, and Biosciences Division supports experimental, theoretical, and computational research to provide fundamental understanding of chemical transformations and energy flow in systems relevant to DOE missions. This knowledge serves as a basis for the development of new processes for the generation, storage, and use of energy and for mitigation of the environmental impacts of energy use.	**71,913,000**
Office of Basic Energy Sciences - Materials Sciences and Engineering Division	The Office of Basic Energy Sciences supports fundamental research to understand, predict, and ultimately control matter and energy at the electronic, atomic, and molecular levels in order to provide the foundations for new energy technologies and to support DOE missions in energy, environment, and national security. The Materials Sciences and Engineering Division supports fundamental experimental and theoretical research to provide the knowledge base for the discovery and design of new materials with novel structures, functions, and properties. This knowledge serves as a basis for the development of new materials for the generation, storage, and use of energy and for mitigation of the environmental impacts of energy use.	**22,645,000**
Energy Frontier Research Centers	Under this initiative, the Office of Science funds centers that conduct basic and advanced discovery research to accelerate advanced energy technologies, including renewable energy technologies. The centers are overseen by program staff in the Materials Sciences and Engineering subprogram and the Chemical Sciences, Geosciences, and Biosciences subprogram.	**38,200,000**

Initiative name and implementing office	Description	FY 2010 and FY 2011 total (actual and estimated) obligations
Energy Innovation Hubs	This initiative is part of a broad-based research strategy to help the United States meet energy and climate challenges. Under this initiative, DOE has launched three Energy Innovation Hubs modeled after its Bioenergy Research Centers. The hubs will help advance highly promising areas of energy science and engineering from the early stage of research to the point where the technology can be handed off to the private sector. The Fuels from Sunlight Hub in particular focuses on renewable energy, seeking to develop an effective solar energy to chemical fuel conversion system.	42,000,000
Total		$2,342,765,827

Source: GAO analysis of agency-provided data.

[a]The Advanced Research Projects Agency-Energy's authorizing legislation, the America Creating Opportunities to Meaningfully Promote Excellence in Technology, Education, and Science Reauthorization Act of 2007, (Pub. L. No. 110-69 (2007)) was reauthorized on January 4, 2011; and funding for the program was authorized through fiscal year 2013.

[b]The Energy Policy Act of 2005 and the Energy Independence and Security Act of 2007 provide authorization for these activities through fiscal year 2015 or 2020 depending upon the activity.

[c]The American Recovery and Reinvestment Act of 2009 added section 1705 to the Energy Policy Act of 2005. This initiative had a sunset date of September 30, 2011.

Table 6: Environmental Protection Agency Solar-Related Initiatives

Initiative name and implementing office	Description	FY 2010 and FY 2011 total (actual and estimated) obligations
Office of Research and Development		
P3: People, Prosperity & the Planet Student Design Competition for Sustainability	This initiative is a competitive grants program in which teams of college or university students propose the R&D of sustainable designs related to water, energy, agriculture, the built environment or materials and chemicals. Selected teams are awarded $15,000 grants to develop and demonstrate their ideas at the annual National Sustainable Design Expo in Washington, D.C., where they compete for a $90,000 grant to take their design to the next level or to deployment. Proposals related to the collection of solar energy and the use of solar energy for drinking water purification have been among those funded.	$200,000
Total		$200,000

Source: GAO analysis of agency-provided data.

Table 7: National Aeronautics and Space Administration Solar-Related Initiatives

Initiative name and implementing office	Description	FY 2010 and FY 2011 total (actual and estimated) obligations
Office of the Chief Technologist		
High Efficiency Space Power Systems	As part of the Exploration Technology Development Program's Enabling Technology Development and Demonstration projects, NASA will develop technologies to provide low-cost, abundant power for deep-space missions, including solar photovoltaic systems. The initiative will also demonstrate dual-use technologies for clean and renewable energy for terrestrial applications.	$308,000
Space Technology Research Grants Program	This is the coupling of activities within two Office of the Chief Technologist programs, the Space Technology Research Grants Program and the NASA Innovative Advance Concepts. The Space Technology Research Grants Program's objective is to accelerate the development of push technologies through innovative efforts with high-risk/high payoff through the execution. The NASA Innovative Advance Concepts program's objective is to provide an initial study of visionary technology concepts that are aerospace architecture, system or mission, exciting by exploring a potential breakthrough, far-term implementing design more than 10 years out and have scientific and engineering credibility.	265,275
Solar Sail Demonstration	The Solar Sail Demonstration will advance the state of the art of the use of solar radiation pressure for in-space propulsion (solar sailing). The demonstration of this technology will enable many high-energy space missions that are not feasible using conventional propulsion technologies.	55,993
Office of the NASA Associate Administrator		
Small Business Innovations Research/Small Business Technology Transfer Research	In these programs, NASA seeks small, high-technology companies to participate in government-sponsored research and development efforts in technology areas critical to NASA's missions. One of the areas of focus is energy conversion, including the development of advanced solar photovoltaic technologies to improve efficiency and reliability of solar power for space exploration missions. Another area of focus is energy generation and the development of renewable sources of energy, including space-based solar power generation and other renewable energy technologies that enable operation over wide temperature ranges and harsh environmental conditions.	31,889,077
Science Mission Directorate		
High Temperature Photovoltaics	This initiative provides support to the Organometallic Vapor Phase Epitaxy Facility to develop solar cells for operation at elevated temperatures.	450,000
Solar Probe Plus Technology Development	This initiative provides performance testing of triple junction solar cells in elevated temperature and radiation environments like that seen near the sun.	350,000[a]
Space Technology Program		
Centennial Challenges	The Centennial Challenges Program seeks innovative solutions to technical problems that can drive progress in aerospace technology of value to NASA's missions in space operations, science, exploration, and aeronautics. The initiative encourages the participation of independent teams, individual inventors, student groups, and private companies in aerospace research and development. Competition topics include solar and other renewable energy technologies.	1,500,000
Total		**$34,818,345**

Source: GAO analysis of agency-provided data.

[a]All or part of the agency-provided obligations are estimated.

Table 8: National Science Foundation Solar-Related Initiatives

Initiative name and implementing office	Description	FY2010 and FY2011 total (actual and estimated) obligations
Directorate for Engineering		
Catalysis and Biocatalysis	The program funds a variety of academic research grants, including those in a research area known as "Catalysts and Studies for Alternative Energy Systems, such as Electro- and Photocatalysis." To meet national needs, the program prioritizes proposals for conversion of biomass to fuels and chemicals, development of alternative energy sources, and transition to green products and processes.	$1,269,911[a]
Emerging Frontiers in Research and Innovation	The initiative funds grants for interdisciplinary engineering research with the potential to create a significant impact or meet national needs. Under the program in fiscal year 2010, NSF coordinated with DOE to call for proposals that help address the need for solar and wind energy storage.	8,006,000
Energy, Power, and Adaptive Systems	The program provides grants for academic researchers to design and study intelligent and adaptive engineering networks with an emphasis on electric power electronics, networks and grids. This focus includes generation, distribution, transmission and integration of renewable energy systems and alternative energy technologies, including solar cells, ocean waves, wind, and hydropower. The program also supports laboratory and curriculum development to integrate research and education.	4,985,617
Energy for Sustainability Program	The program funds fundamental academic research and education activities through grants to enable innovative processes for the sustainable production of electricity and transportation fuels. The program emphasizes two primary fuel sources— biofuels/bioenergy and photovoltaic solar energy—and also supports research in wind and wave energy, sustainable energy technology assessment, and fuel cells.	7,160,607
Engineering Research Centers	Since 1985, the NSF Engineering Research Center program has funded interdisciplinary academic-led centers that create and sustain an integrated research environment to advance fundamental engineering knowledge and engineered systems. In fiscal year 2010, the program issued a solicitation that included a call for energy-related proposals. As a result of that call, two new renewable energy-related engineering research centers—one in solar photovoltaic systems and one in electrical energy transmission–will be started with co-funding from DOE. In addition in fiscal year 2010, of the 15 engineering research centers active at that time, there were two in the renewable energy area—one that conducted biofuel research and one that conducted "smart grid" electrical energy distribution research.	1,502,036
Directorate for Mathematical and Physical Sciences		
CHE-DMR-DMS Solar Energy Initiative (SOLAR)	The initiative provides grants to support interdisciplinary efforts by research groups—each of which includes a chemist, a materials researcher, and a mathematician or statistician—to address the scientific challenges of highly efficient harvesting, conversion, and storage of solar energy.	22,165,125
Chemical Catalysis	The program provides grants for experimental and theoretical academic research on chemical agents known as catalysts and pre-catalysts. Specifically, the program supports research on topics such as solar energy conversion, biomass conversion, and other energy-related processes.	3,672,260
Total		$48,761,556

Source: GAO analysis of agency-provided data.

[a]All or part of the agency-provided obligations are estimated.

Table 9: U.S. Department of Agriculture Solar-Related Initiatives

Initiative name and implementing office	Description	FY2010 and FY2011 total (actual and estimated) obligations
Agricultural Research Service		
Bioenergy National Program	The mission of this Agricultural Research Service research program is to develop technologies for the sustainable commercial production of biofuels by the agricultural sector in ways that enhance natural resources without disrupting existing food, feed, and fiber markets. Agricultural Research Service efforts under this initiative include research on new varieties and hybrids of bioenergy feedstocks, practices and systems that maximize the yield of bioenergy feedstocks, and commercially preferred biorefining technologies.	$270,000
National Institute of Food and Agriculture		
Sustainable Agriculture Research and Education Program	This regionally administered competitive grant program supports research and education projects related to sustainable agriculture. It also includes a professional development program to provide education and training on sustainable agriculture to agricultural professionals. It does not specifically call for renewable energy proposals but has funded projects that include renewable energy elements, such as alternative biomass production, on-farm biomass conversion methods, and small-scale methane digesters.	107,247
Tribal Colleges Research Grants Program	This program is a competitive grant program supporting fundamental and/or applied agricultural research projects that address high-priority food and agricultural science concerns of tribal, national, or multistate significance. Its mission is to encourage the engagement and participation of Native American students in food and agricultural sciences research through experiential learning. Program priorities are based on National Institute of Food Agriculture's national critical needs areas, which include development of sustainable energy, among others.	200,000
Total		$577,247

Source: GAO analysis of agency-provided data.

Appendix III: GAO's Questionnaire for Federal Agencies with Initiatives Supporting Solar Energy Technologies

 **G A O**

United States Government Accountability Office

QUESTIONS ABOUT FEDERAL SOLAR-RELATED INITIATIVE:
[PRE-POPULATED INFORMATION]

Introduction

The United States Government Accountability Office (GAO), an independent, legislative branch agency, is examining how the federal government supports research and development, and commercialization of solar energy technologies through solar-related initiatives (GAO job code 361329). Solar-related initiatives are those that could or do support solar energy research and/or technology, either exclusively or as part of a broader initiative. GAO is undertaking this work at the request of the Chairman of the House Committee on Science, Space, and Technology. This work has three objectives: (1) identify the key characteristics of solar energy initiatives supported by federal agencies, (2) determine the extent of any potential fragmentation, overlap, or duplication in federal solar-related initiatives, and (3) examine the extent to which federal agencies coordinate to achieve common goals for their solar-related initiatives.

In 2011, we asked your agency to provide us with information and data on the renewable energy initiatives that were ongoing at your agency in fiscal year 2010 (GAO job code 361185). Using that information, we have identified any initiatives that your agency indicated had a solar component. Through this supplemental questionnaire, we are asking you to provide additional information about the initiative that you identified for us. After we receive your response, we will follow up, to discuss and clarify any outstanding questions about the initiative as they relate to the objectives stated above.

This questionnaire is divided into four sections and should be completed by agency officials involved with the [Pre-populated information]

1) SECTION I – General Information About Initiative
2) SECTION II – Funding for Initiative
3) SECTION III – Coordination
4) SECTION IV – Other Information

Please complete all four sections and return it to one of the individuals mentioned below by **March xx, 2012.** When returning the questionnaire, please attach any relevant supporting documentation to your email.

If you have any questions or comments about this questionnaire, please call or e-mail Holly Sasso at (202) 512-4888 (sassoh@gao.gov) or Tanya Doriss at (202) 512-9546 (dorisst@gao.gov). Thank you very much for your assistance.

Page 1 of 10

QUESTIONS ABOUT FEDERAL SOLAR INITIATIVES   [PRE-POPULATED INFORMATION]

Instructions

This questionnaire can be completed using MS-Word and returned via e-mail to sassoh@gao.gov
or dorisst@gao.gov. Please complete this questionnaire and return it by **March xx, 2012**.

1) Use your mouse to navigate by clicking on the field or check box ☐ you wish to answer.

2) To select a check box or button, click on the center of the box, and an 'X' will appear.

3) To deselect a check box response, click on the center of the box, and the 'X' will
    disappear.

4) To answer a question that requires a comment, click on the answer box   and begin
    typing. The box will expand to accommodate your answer.

Page 2 of 10

QUESTIONS ABOUT FEDERAL SOLAR INITIATIVES   [PRE-POPULATED INFORMATION]

SECTION I: GENERAL INFORMATION ABOUT INITIATIVE

In 2011, we compiled preliminary information provided by your agency for this solar-related initiative as part of our renewable energy initiatives report (GAO job code 361185). We define solar-related initiatives as those that could or do support solar energy research and/or technology, either exclusively or as part of a broader initiative. The initiative itself, may support solely, or in part, solar *research and development* — including efforts ranging from defining scientific concepts to those applying and demonstrating new and improved technologies. The initiative may also support solely, or in part, solar technology *commercialization*—such as, efforts that bridge the gap between research and development activities and the marketplace by transitioning technologies to commercial applications. We have excluded from our review any solar-related initiatives that solely support deployment — for example, efforts to facilitate or achieve widespread use of technologies either in the commercial market or for nonmarket uses such as defense, through their construction, operation, or use.

Your agency provided the following information about the initiative to a previous GAO job (job code 361185).

Initiative Title:	[Pre-populated information]
Implementing Agency:	[Pre-populated information]
Implementing Office:	[Pre-populated information]
Initiative Description:	[Pre-populated information]
Type of Solar Technology Supported:	[Pre-populated information]
Funding Mechanism:	[Pre-populated information]
Recipient Type:	[Pre-populated information]

1. **Is the information provided above about the initiative correct and complete?** *We define an initiative as programmatic or mission-area activities, or a group of agency activities serving similar purposes or functions. We are not considering individual projects within an initiative — such as a specific grant award — to be initiatives in themselves.*

    Yes .... ☐
    No .... ☐

    If you answered No, what information is incorrect or incomplete, and what is the correct or complete information? Please use the space below.

QUESTIONS ABOUT FEDERAL SOLAR INITIATIVES   [PRE-POPULATED INFORMATION]

2. Your agency previously indicated to GAO that this initiative was active in fiscal year 2010.
    **Please confirm that this initiative was active—for either solar-related OR any other
    activities—at your agency at any time during fiscal year 2010.** *We consider an initiative
    to be "active" if the initiative was planned or funded or implemented or authorized in the
    fiscal year described. We are interested in those initiatives for which funds were obligated in
    a previous fiscal year, and for which activities were still on-going in fiscal year 2010.*

    Was this initiative active at your agency at any time during fiscal year 2010?

    Yes ....☐
    No............................☐

    If you answered No, please explain in the space below.

3. **Was this initiative active— for either solar OR any other activities—at your agency at
    any time during fiscal year 2011?**

    Yes ....☐
    No............................☐

    If you answered No, please explain in the space below. *For example, if the initiative was
    defunded.*

4. **Please describe your initiative's goals in the space below.**

5. **Does your initiative have goals that are <u>directly related to solar energy research and/or
    technology</u>?**

    Yes ....☐
    No............................☐

    If you answered Yes, please describe the goals for solar energy research and/or technology in
    the space below. Please attach a copy of, or provide weblinks in the space below, to relevant
    documentation, such as strategic plans, annual reports, program plans, grant solicitations,
    funding opportunity announcements, or requests for proposals:

    Page 4 of 10

QUESTIONS ABOUT FEDERAL SOLAR INITIATIVES   [PRE-POPULATED INFORMATION]

6. We understand that each agency defines technology advancement activities in a unique manner; however, in order to compare activities across agencies, we have developed definitions that can be applied broadly and have listed them below. Please determine which definition(s) best fit your individual initiative and answer for all of the initiative's activities, if applicable, not only those that are solar-related.

Which technology advancement activity(ies) apply to this initiative? Please check Yes or No for each activity.	Yes ▼	No ▼
a. **Basic research?** .... *Basic research includes efforts to explore and define scientific or engineering concepts, or is conducted to investigate the nature of a subject without targeting any specific technology.*	☐	☐
b. **Applied research?** .... *Applied research includes efforts to develop new scientific or engineering knowledge to create new and improved technologies.*	☐	☐
c. **Demonstration activities?** .... *Demonstration activities include efforts to operate new or improved technologies to collect information on their performance and assess readiness for widespread use.*	☐	☐
d. **Commercialization?** .... *Commercialization includes efforts to bridge the gap between research and demonstration activities, and venture capital funding and marketing activities, through transitioning technologies to commercial applications.*	☐	☐
e. **Deployment?** .... *Deployment includes efforts to facilitate or achieve widespread use of technologies either in the commercial market or for nonmarket uses such as defense, through their construction, operation or use.*	☐	☐
f. **Other?** *Please specify in space below*	☐	☐

Page 5 of 10

QUESTIONS ABOUT FEDERAL SOLAR INITIATIVES – [PRE-POPULATED INFORMATION]

SECTION II: FUNDING FOR INITIATIVE

We are interested in how much money goes toward the solar-related activities your initiative supports. We would like to be able to report obligations data across all federal initiatives for fiscal years 2010 and 2011. If necessary, please consult with staff in your agency's budget office to answer these funding-related questions.

Note: We may request copies of supporting documentation for the numbers you provide below.

For questions in this section, please use the following definition: *An **obligation** is a definite commitment that creates a legal liability of the government for the payment of goods and services ordered or received, or a legal duty on the part of the United States that could mature into a legal liability. Payment may be made immediately or in the future. An agency incurs an obligation, for example, when it places an order, signs a contract, awards a grant, purchases a service, or takes other actions that require the government to make payments to the public or from one government account to another.*

7. **For fiscal years 2010 and 2011, please provide actual or estimated obligations data for [Pre-populated information]'s solar activities only.** We understand that some agencies may not be able to provide actual or estimated obligations for solar-related activities.

	Total obligations for initiative's **solar activities**	Check if the total is an **estimate**	Check if you are unable to provide obligations
a. Fiscal year 2010?................	$_____	☐	☐
b. Fiscal year 2011?................	$_____	☐	☐

c. **If the total is an estimate, please describe how the estimate was determined and describe any limitations associated with this estimate.** *(If the total was not an estimate, check N/A and proceed to question 8.)*

FY2010:

FY2011:

N/A: ☐

d. **If you were unable to provide actual or estimated obligations, please explain why, in the space below.** *(If you did provide obligations data, check N/A and proceed to question 8.)*

FY2010:

FY2011:

N/A: ☐

Page 6 of 10

QUESTIONS ABOUT FEDERAL SOLAR INITIATIVES [PRE-POPULATED INFORMATION]

Questions 8 through 10 ask about specific projects funded or supported by your initiative in fiscal years 2010 and 2011.

8. **How many total projects were active in this initiative in fiscal years 2010 and 2011?** *We consider individual projects to exist as part of an overall initiative—for example, specific grant awards, or agreements, or contracts that are supported by your initiative. We consider an initiative to be "active" if the initiative was planned or funded or implemented or authorized in the fiscal year described.*

FY2010:

FY2011:

9. **Were all projects listed in question 8 solar-related?**

Yes .... ☐

No .... ☐

If you answered No, how many of the projects listed above were solar-related in fiscal years 2010 and 2011?

FY2010:

FY2011:

10. As part of our reporting to Congress, we'd like to provide a few examples of projects that have been funded. **In 800 or fewer characters, please briefly describe two or three examples of specific solar-related projects funded or supported by this initiative in the space below:**

     a. Project #1

     b. Project #2

     c. Project #3

11. **If needed, please use the space below to provide further details about your answers in Section II of this questionnaire.**

Page 7 of 10

QUESTIONS ABOUT FEDERAL SOLAR INITIATIVES   [PRE-POPULATED INFORMATION]

SECTION III: COORDINATION

The Congress and GAO are interested in understanding how agencies coordinate either within the agency or with other agencies on solar-related initiatives.

12. Do you or staff from this initiative coordinate with other solar-related technology initiatives within your agency or with other agencies?

Yes ....☐
No.............................☐

If you answered Yes to the above question, please provide up to five examples in the table below of how you coordinate within your agency or with other agencies.

Name/Title of Coordination Activities	Description of Coordination Activities	Does this Coordination Activity Involve Coordinating with Other Agencies?	
		Yes ▼	No ▼
1.		☐	☐
2.		☐	☐
3.		☐	☐
4.		☐	☐
5.		☐	☐

13. We are interested in the extent to which agencies make efforts to ensure that their initiatives do not overlap or duplicate efforts made by other agencies. **Does your office or agency collect information on solar initiatives that are active in other offices or agencies in order to assess potential overlap or duplication?**

Yes ....☐
No.............................☐

If you answered Yes, please explain how your office or agency collects the information and how your agency assesses the information.

Page 8 of 10

14. **When reviewing recipient requests for funding, do you determine whether or not they have received other sources of federal funding for the project?**

  ┌─ Yes ....☐
  │ No.............................☐
  ↓

  If you answered Yes, please describe in the space below what information you review and how you determine whether or not they have received other sources of federal funding for the project.

15. **Does your initiative have any policies that either prohibit or permit recipients from receiving other sources of federal funding for the project your initiative supports?**

  ┌─ Yes ....☐
  ├─ No.............................☐
  ↓

  If you answered Yes, please explain the policy in the space below. Please also attach a copy of the policy to this questionnaire or provide weblinks to the policy on your agency's website.

  If you answered No, and wish to provide further information, please do so in the space below.

QUESTIONS ABOUT FEDERAL SOLAR INITIATIVES   [PRE-POPULATED INFORMATION]

SECTION IV: OTHER INFORMATION

16. **Are there any additional data, nuances, further sources of information, or comments
not covered previously that would help us further understand and report on how this
initiative is being implemented? If so, please describe and/or provide link(s) to the
relevant website(s) in the space below.** *If you do not have additional information to
provide, it is appropriate to leave this question blank.* If applicable, please provide electronic
copies of the relevant documents (e.g., strategic planning documents, performance metrics,
RFPs, regulatory provisions, etc.) when returning this questionnaire. Alternatively, you can
suggest that further discussion about this initiative be conducted through follow-up
conversations with GAO.

17. **Please provide the contact information for a representative from your agency for any
follow-up questions we may have about this questionnaire:**

   Name:

   Title:

   Agency:

   Office Division:

   Email:

   Phone:

Please remember to attach any relevant supporting documentation when returning this
questionnaire to either or sassoh@gao.gov or dorisst@gao.gov.

Thank you for your time!

Page 10 of 10

Appendix IV: GAO Contact and Staff Acknowledgments

GAO Contact	Frank Rusco, (202) 512-3841 or ruscof@gao.gov
Staff Acknowledgments	In addition to the individual named above, key contributors to this report included Karla Springer (Assistant Director), Tanya Doriss, Cindy Gilbert, Jessica Lemke, Cynthia Norris, Jerome Sandau, Holly Sasso, Maria Stattel, and Barbara Timmerman.

GAO's Mission	The Government Accountability Office, the audit, evaluation, and investigative arm of Congress, exists to support Congress in meeting its constitutional responsibilities and to help improve the performance and accountability of the federal government for the American people. GAO examines the use of public funds; evaluates federal programs and policies; and provides analyses, recommendations, and other assistance to help Congress make informed oversight, policy, and funding decisions. GAO's commitment to good government is reflected in its core values of accountability, integrity, and reliability.
Obtaining Copies of GAO Reports and Testimony	The fastest and easiest way to obtain copies of GAO documents at no cost is through GAO's website (www.gao.gov). Each weekday afternoon, GAO posts on its website newly released reports, testimony, and correspondence. To have GAO e-mail you a list of newly posted products, go to www.gao.gov and select "E-mail Updates."
Order by Phone	The price of each GAO publication reflects GAO's actual cost of production and distribution and depends on the number of pages in the publication and whether the publication is printed in color or black and white. Pricing and ordering information is posted on GAO's website, http://www.gao.gov/ordering.htm.
	Place orders by calling (202) 512-6000, toll free (866) 801-7077, or TDD (202) 512-2537.
	Orders may be paid for using American Express, Discover Card, MasterCard, Visa, check, or money order. Call for additional information.
Connect with GAO	Connect with GAO on Facebook, Flickr, Twitter, and YouTube. Subscribe to our RSS Feeds or E-mail Updates. Listen to our Podcasts. Visit GAO on the web at www.gao.gov.
To Report Fraud, Waste, and Abuse in Federal Programs	Contact:
	Website: www.gao.gov/fraudnet/fraudnet.htm E-mail: fraudnet@gao.gov Automated answering system: (800) 424-5454 or (202) 512-7470
Congressional Relations	Katherine Siggerud, Managing Director, siggerudk@gao.gov, (202) 512-4400, U.S. Government Accountability Office, 441 G Street NW, Room 7125, Washington, DC 20548
Public Affairs	Chuck Young, Managing Director, youngc1@gao.gov, (202) 512-4800 U.S. Government Accountability Office, 441 G Street NW, Room 7149 Washington, DC 20548

Please Print on Recycled Paper.